T0226112

Advances in 21st Century Human Settlements

Series Editor

Bharat Dahiya, School of Global Studies, Thammasat University, Bangkok, Thailand

Editorial Board

Andrew Kirby, Arizona State University, Tempe, USA

Erhard Friedberg, Sciences Po-Paris, France

Rana P. B. Singh, Banaras Hindu University, Varanasi, India

Kongjian Yu, Peking University, Beijing, China

Mohamed El Sioufi, Monash University, Australia

Tim Campbell, Woodrow Wilson Center, USA

Yoshitsugu Hayashi, Chubu University, Kasugai, Japan

Xuemei Bai, Australian National University, Australia

Dagmar Haase, Humboldt University, Germany

Ben C. Arimah, United Nations Human Settlements Program, Nairobi, Kenya

Indexed by SCOPUS

This Series focuses on the entire spectrum of human settlements – from rural to urban, in different regions of the world, with questions such as: What factors cause and guide the process of change in human settlements from rural to urban in character, from hamlets and villages to towns, cities and megacities? Is this process different across time and space, how and why? Is there a future for rural life? Is it possible or not to have industrial development in rural settlements, and how? Why does 'urban shrinkage' occur? Are the rural areas urbanizing or is that urban areas are undergoing 'ruralisation' (in form of underserviced slums)? What are the challenges faced by 'mega urban regions', and how they can be/are being addressed? What drives economic dynamism in human settlements? Is the urban-based economic growth paradigm the only answer to the quest for sustainable development, or is there an urgent need to balance between economic growth on one hand and ecosystem restoration and conservation on the other – for the future sustainability of human habitats? How and what new technology is helping to achieve sustainable development in human settlements? What sort of changes in the current planning, management and governance of human settlements are needed to face the changing environment including the climate and increasing disaster risks? What is the uniqueness of the new 'socio-cultural spaces' that emerge in human settlements, and how they change over time? As rural settlements become urban, are the new 'urban spaces' resulting in the loss of rural life and 'socio-cultural spaces'? What is leading the preservation of rural 'socio-cultural spaces' within the urbanizing world, and how? What is the emerging nature of the rural-urban interface, and what factors influence it? What are the emerging perspectives that help understand the human-environment-culture complex through the study of human settlements and the related ecosystems, and how do they transform our understanding of cultural landscapes and 'waterscapes' in the 21st Century? What else is and/or likely to be new vis-à-vis human settlements – now and in the future? The Series, therefore, welcomes contributions with fresh cognitive perspectives to understand the new and emerging realities of the 21st Century human settlements. Such perspectives will include a multidisciplinary analysis, constituting of the demographic, spatio-economic, environmental, technological, and planning, management and governance lenses.

If you are interested in submitting a proposal for this series, please contact the Series Editor, or the Publishing Editor:
Bharat Dahiya (bharatdahiya@gmail.com) or
Loyola D'Silva (loyola.dsilva@springer.com)

More information about this series at https://link.springer.com/bookseries/13196

Sandeep Narayan Kundu
Editor

Geospatial Data Analytics and Urban Applications

 Springer

Editor
Sandeep Narayan Kundu
Hydrography and Geoconsulting
FUGRO
Singapore, Singapore

ISSN 2198-2546   ISSN 2198-2554 (electronic)
Advances in 21st Century Human Settlements
ISBN 978-981-16-7651-2   ISBN 978-981-16-7649-9 (eBook)
https://doi.org/10.1007/978-981-16-7649-9

© The Editor(s) (if applicable) and The Author(s), under exclusive license to Springer Nature Singapore Pte Ltd. 2022
This work is subject to copyright. All rights are solely and exclusively licensed by the Publisher, whether the whole or part of the material is concerned, specifically the rights of translation, reprinting, reuse of illustrations, recitation, broadcasting, reproduction on microfilms or in any other physical way, and transmission or information storage and retrieval, electronic adaptation, computer software, or by similar or dissimilar methodology now known or hereafter developed.
The use of general descriptive names, registered names, trademarks, service marks, etc. in this publication does not imply, even in the absence of a specific statement, that such names are exempt from the relevant protective laws and regulations and therefore free for general use.
The publisher, the authors and the editors are safe to assume that the advice and information in this book are believed to be true and accurate at the date of publication. Neither the publisher nor the authors or the editors give a warranty, expressed or implied, with respect to the material contained herein or for any errors or omissions that may have been made. The publisher remains neutral with regard to jurisdictional claims in published maps and institutional affiliations.

This Springer imprint is published by the registered company Springer Nature Singapore Pte Ltd.
The registered company address is: 152 Beach Road, #21-01/04 Gateway East, Singapore 189721, Singapore

I dedicate this book to all spatial scientists who are working to make our world a safe and liveable place by contributing insights to sustainability through resource optimization.

Preface

In today's digital world, human activities are more interconnected and impact the way we use our limited resources. In dense urban setups, it is more imperative that resource utilization is optimized, and sustainable. Conventional paradigms have limitations in connecting the missing links between the cascading effects of unsustainable means of resource utilization. As more and more data are being recorded for every activity and made publicly available for data scientists, insights can now be drawn with multiple perspectives to identify patterns and trends. However, there are several challenges. The first one is the selection of the right dataset from the myriad volume of Big data available to us today. Next challenge is the selection of the right algorithm on which the analytics are to be founded on top of this data. The final challenge lies in the interpretation of the results and this defines the insights and the actions to be taken by authorities for solving problems. These three challenges constitute the essential components of a discipline which is now known as Data Science.

Data is nothing but factual instances collected over time on any natural or anthropogenic activity. Digital data is data stored in computers with some sort of structure. Spatial data os data with a reference to geographical location. In essence, all data contain some spatial information. Weather, for example, is spatial data on temperature, rainfall and wind at a location. Similarly, motion of an object, like a car, is spatial data as it contains spatial information on the locations it traverses through. Even social media posts from a mobile device are spatial in nature, as the location of the device is recorded in the metadata. Therefore, spatial data analytics hold a special place in data science where the insights are drawn with a geographical connotation. This is particularly important for urban settings.

Urban Settings pose several problems most of which are related to supply chain issues. Sustainability and continuity of usual activities are easily impacted because of any disruptions and these disruptions have the potential to cascade into major catastrophic events impacting lives and livelihood. Spatial data science, therefore, has a major role to play in addressing these urban issues. This is why spatial data is being collected over time from almost all urban activities. The analytics on spatial data not only help us understand the state of health of several aspects of urban life but also support informed decisions on future policies and pathways.

This book is a collection of work of emerging spatial data scientists who experimented with temporal subsets of data from multiple Big data portals and used them for identifying and solving urban issues. The datasets were spatialized as needed and then were analyzed using spatial statistical techniques for insightful interpretation of the patterns obtained. The patterns found were influenced by a combination of anthropogenic and natural factors. However, the spatial analytics brought forth the prime drivers which greatly helps us understand the trends for the future thereby presenting us with an opportunity to favourably influence the potential adverse outcomes.

The chapter "Spatial Big Data and Urban Analytics: An Introduction" is an introduction to Spatial Data Science and its applications. It outlines the stages in a data science project and emphasizes on its application, especially in urban settings, for maintaining a balance for sustaining life and modern activities. The chapter "Preferential Home Search in Urban Settings" is a study on people's considerations for searching a livable place in Singapore. It demonstrates how spatial data can help them refine their search based on individual priorities. Major property portals have basic spatial criteria to help home seekers, but this chapter identifies areas where those can be enhanced. The chapter "Air Quality Dynamics and Urban Heat Island Effects During COVID-19" compares a COVID-19 setting against a pre-COVID-19 setting to understand air quality dynamic and urban heat island effects in both Wuhan City and New York City. The study compares patterns in the two distant geographies. The chapter "Mass Rapid Transit and Population Dynamics During Covid-19 in Singapore" studies Mass Rapid Transit (MRT) ridership in Singapore and compares weekly and weekend travel patterns amid COVID-19 lockdown and after lockdown restrictions were partially lifted. The chapter "A Geospatial Analysis of Tweets During Post-circuit Breaker in Singapore" demonstrates how spatial data analytics can be applied to Twitter data for understanding the relationship between service amenities and land use in urban setups. It is worth mentioning that tweets and other social media posts are now being increasingly used by governmental agencies for assessment of public sentiment on a myriad of issues. The chapter "Space-Time Analytics of New York City Shooting Incidents" investigates the space-time analytics of the New York Shooting which has an impact on policing of administrative units to address the shooting incidents. The chapter "Spatial and Temporal Patterns of Tourist Source Market Emissiveness: A Study of Shanghai, China" deals on Tourism emissiveness in China with Shanghai being the focus of the study. The chapter "Spatial Perspectives of Crime Patterns in Chicago Amid Covid-19" studies the crime patterns in Chicago and reflects on how those patterns change with the onset of COVID-related restrictions on anthropogenic activities. The chapter "Geospatial Analysis of Grab Trips in Singapore" evaluates the use of private car hires in Singapore. This space and time study is based on data derived from the major hire car aggregator Grab. The final chapter, the chapter "Ecological Vulnerability of Nyingchi, Tibet", is a study on spatial variability of ecological vulnerability in the city of Nyinchi in Tibet.

I hope the book shall be a good reference for spatial data scientists who can leverage on the broad applications of spatial data science in several urban settings that is presented. It is envisaged that the book shall open windows for advancing insightful applications as availability and access of spatial data is only going to improve in future.

Singapore, Singapore

Dr. Sandeep Narayan Kundu

Acknowledgements

I thank all the contributors, who despite their busy academic life, could find time to collaborate contribute to their chapters. Each of their individual perspective on the subject complemented the data science project on the course which has shaped the respective chapters.

I also would like to thank my wife Priyanka, daughter Shreya and son Shakti for allowing me to work on weekends for the completion of this book.

Finally, this dedication will not be complete without the mention of my father Dr. Manmath Kundu and my mother Dr. Binodini Patra, who have a stellar role in inculcating a strong academic and scientific mindset in me.

Dr. Sandeep Narayan Kundu

Contents

Contributors

Xiang Jing Ang Faculty of Arts and Social Sciences, National University of Singapore, Singapore, Singapore

Xi Bohao Pudong, Shanghai, China

Yan ChengCheng National University of Singapore, Singapore, Singapore

Huang Fengjue Arts Link, National University of Singapore, Singapore, Singapore

Mark Frank Ratnam National University of Singapore, Singapore, Singapore

Chee Young Goh Faculty of Arts and Social Sciences, National University of Singapore, Singapore, Singapore

Hu Guanxian Arts Link, National University of Singapore, Singapore, Singapore

Zou Hongyi Singapore, Singapore

Wang Jifei National University of Singapore, Kent Ridge, Singapore

Wang Jing Singapore, Singapore

Jia Jingnan Pudong, Shanghai, China

Soomin Kang National University of Singapore, Singapore, Singapore

Sandeep Narayan Kundu Fugro Singapore Marine, Ballota Park, Singapore

Dina Labiba National University of Singapore, Singapore, Singapore

Luo LaiWen National University of Singapore, Singapore, Singapore

Zhang Liqing National University of Singapore, Singapore, Singapore

Sherie Loh Wei Singapore, Singapore

Sharon Low National University of Singapore, Singapore, Singapore

Dini Aprilia Norvyani National University of Singapore, Singapore, Singapore

Gu Qianhua Singapore, Singapore

Qi RuiYuan National University of Singapore, Singapore, Singapore

Li Ruoyu Changchun, China

Sun Tong National University of Singapore, Kent Ridge, Singapore

Lei Wang National University of Singapore, Singapore, Singapore

Hui Ling Wee Faculty of Arts and Social Sciences, National University of Singapore, Singapore, Singapore

Liu Weiyu National University of Singapore, Kent Ridge, Singapore

Lyu Wenling Pudong, Shanghai, China

Zhu Wenzhe Arts Link, National University of Singapore, Singapore, Singapore

Ji Xin Arts Link, National University of Singapore, Singapore, Singapore

Shuhan Yang National University of Singapore, Singapore, Singapore

Phang Yong Xin Singapore, Singapore

Xu Yuanyuan National University of Singapore, Kent Ridge, Singapore

Wang YunJie National University of Singapore, Singapore, Singapore

Xu Yuting Hindhede Walk, Singapore, Singapore

Yingzhe Zhang Faculty of Arts and Social Sciences, National University of Singapore, Singapore, Singapore

Lim Zhu An Singapore, Singapore

Wang Ziwen Pudong, Shanghai, China

Abbreviations

AI	Artificial Intelligence
API	Application Processing Interface
AQI	Air Quality Index
CAAQS	Canadian Ambient Air Quality Standards
CBD	Central Business District
CHAS	Community Health Assist Scheme
CNN	Cable News Network
COVID-19	Coronavirus Disease-2019
CSCL	City Street Centre Line
DUS	Dual-Use Scheme
EHSA	Emerging
EPA	Environmental Protection Agency
EPSG	European Petroleum Survey Group
ESRI	Environmental Systems Research Institute
EVI	Ecological Vulnerability Index
EVSI	Ecological Vulnerability Standardization Index
GIS	Geographical Information System
GPS	Global Positioning System
HDB	Housing Development Board of Singapore
ICT	Information and Communications Technology
IDW	Inverse Distance Weighted
IoT	Internet of Things
IT	Information Technology
KDE	Kernel Density Estimation
LISA	Local Indicators of Spatial Autocorrelation
LOA	Local Outlier Analysis
LST	Land Surface Temperature
LTA	Land Transportation Authority of Singapore
MGWR	Multi-scale Geographical Weighted Regression
MOE	Ministry of Education
MRT	Mass Rapid Transit

NTA	Neighbourhood Tabulation Areas
NUS	National University of Singapore
NYC	New York City
NYPD	New York Police Department
OLI	Operational Land Imager
OLS	Ordinary Least Squares
PM	Particulate Matter
POI	Points of Interest
PRC	People's Republic of China
PSA	Police Service Area
PSR	Pressure-State-Response
PV	Passenger Volume
SPCA	Spatial Principal Component Analysis
STC	Space-Time Cube
SVM	Support Vector Machine
TIRS	Therman Infrared Sensor
TND	Traditional Neighbourhood Development
TOD	Transit-Oriented Development
UHI	Urban Heat Island
USGS	United States Geological Survey
UTM	Universal Transverse Mercator
VIF	Variation Inflation Factors
VOC	Volatile Organic Compounds

Spatial Big Data and Urban Analytics: An Introduction

Sandeep Narayan Kundu

Abstract Big data refers to the large volumes of data collected by several sensors to monitor natural or anthropogenic activities. These data are temporal in nature and come in various forms, volume, and frequencies. Most of these have a geographical connotation and hence are essentially considered as spatial big data. Treating big data statistically can derive several insights. But spatial analytics add another dimension to the insights which make it critical for decision making. In this chapter, the fundamentals of spatial big data and analytics are discussed from a data science perspective. It also discusses the urban applications of spatial big data analytics.

1 Introduction

Big data refers to the large volumes of data which are being generated on a daily basis. They may come from sensors which monitor urban activities. These data are temporal in nature and come in various forms, volume, and frequencies. The term 'Big Data' was coined to address the problem of storing and handing data which exceed the capacity of present-day computing systems. Innovative spatial data infrastructures are currently being advanced to handle big data. These address connected storage, methods of databasing and mechanisms to efficiently access, query and extract for insightful analytics.

Spatial tools add another dimension by equipping data scientists to gain invaluable spatial insights for decision making. Spatial statistics adds a geographical perspective on top of conventional statistics. Machine learning and Artificial Intelligence are heavily applied for automating the analytics. But the striking aspect of spatial analytics is the geographical visualization of the outcomes. This makes the analysis more intuitive as a picture speaks a thousand words. In spatial data analytics, the foundational database technology still remains unchanged although some tweaks are done to handle spatially enable data. The most critical tool for a spatial data scientists

S. N. Kundu (✉)
Fugro Singapore Marine, 158 Mariam Way, Ballota Park 507083, Singapore

© The Author(s), under exclusive license to Springer Nature Singapore Pte Ltd. 2022
S. N. Kundu (ed.), *Geospatial Data Analytics and Urban Applications*,
Advances in 21st Century Human Settlements,
https://doi.org/10.1007/978-981-16-7649-9_1

is geographic visualization which adds a geographical perspective to conventional analytics.

Some examples of spatial data are raster data (satellite images), vector data (point, line and polygons) and graph data (network data mapped using topology). Spatial data are special as they have a geometry and can also be governed by topological rules. Spatial big data are the above data in huge volume, variety and velocity which come from multiple sources like satellites, drones, vehicles, GPS connected devices and camera etc. In today's work a significant part of big data is essentially spatial big data.

2 Global Challenges and Spatial Data

There are multitude of problems currently plaguing the world. We are faced with challenges to make the world a sustainable and livable place (Fig. 1). Population and demand for resources like land, water and energy are perpetually on the rise. Increased resource utilization is releasing more greenhouse gases and this cascades into catastrophic climate related issues. With the availability of Big Data, spatial data

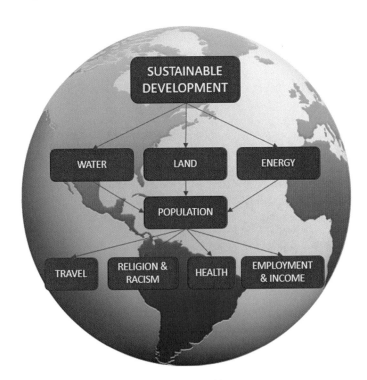

Fig. 1 Sustainable Development issues in today's world

analytics is the key to design and implement sustainable perspectives in addressing today's problems.

The global problems are more amplified in urban settings as activities in each city run in a very interlinked manner. A disruption in one aspect shall have domino effects leading to multiple disruptions over time. In city states like Singapore, the nexus between Land Water and Energy is key to building a liveable world for its residents. To add to this are the impacts of global scale events, like the COVID-19 pandemic and climate change to which the city is equally exposed to.

Over the last 3 decades, urban countries have frameworks for collecting data over time for monitoring its activities. An example is the network of satellites which gather images of earth. Analytics are performed on these multi-temporal, multi-spectral images to extract information on land use, weather, air quality etc. On the ground, we have GPS-based locational data from sensors that track activities of man and machines. Virtually almost all anthropogenic activities are being logged into such location aware devices. The benefits of doing so far surpasses the costs of deployment as they can be actively used for regulatory compliance and governance of the activities of urban life. Policing and crime investigations are also supported by data from such tracking and logging devices. During Covid-19, such locational data sensors are, in fact, helping authorities connect the infected with the likely (close contacts) and support precautionary quarantining for preventing disease spread. Businesses also collect spatial data for understanding their product usage and for addressing newer markets. They target potential customers from their internet search activity and also use loyalty cards of existing customers to monitor their spending patterns and to enhance customer satisfaction. Banks are tracing the location of debit and credit card usage of customers to flag fraudulent transactions.

In essence, most agencies, both government and private, are continuously collecting locational data. This is collected and stored in proprietary systems to support their objectived i.e., gaining insights and extracting business value. Many have started sharing historical data through online portals to benifit third parties. Shared data has sensitive information removed in compliance with data protection acts and some are available for purchase too. With time, more and more data shall be make available for spatial data scientists. As these data are seldom used across multiple agencies for addressing specific urban problems, spatial researchers are perhaps the only species who are now increasingly using them for holistic understanding of urban issues.

3 Spatial Big Data Science

Spatial Big Data project involves several essential components. Apart from Spatial Big Data, the below tools are critical to the success of a project.

- Statistical Capabilities
- Machine Learning or Artificial Intelligence integration

- Database technology
- Collaborative framework (IT infrastructure and Network)
- Scripting (programming capability)
- Visualization Techniques
- Reporting Capabilities

Therefore, selection of a computing environment is key to scale the project to the needs of the analytics. It is most likely that one program does not provide all the needed capabilities. Therefore complementing programs are identified, adopted and integrated into the programming framework.

3.1 Workflow

A data science project broadly follows a workflow, a generic version of which is provided in Fig. 2. The workflow is iteratively adopted to support continuous improvement, which is essential to all data science projects as it's aim is to provide insights continuously for the organization. The starting point is to gain access to the data and prepare it for the study. This involves visiting the data portals of multiple agencies and subsequently conducting an assessment on its appropriateness and usability of the available data. Big data is collected and stored by various agencies for an original purpose. However, a data scientist extracts these data for use

Fig. 2 Generalized spatial data science workflow

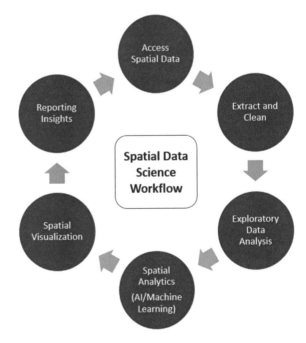

for a particular problem which is different to this original purpose. Therefore, an exploratory data analysis is conducted to check on the utility of the attributes and its completeness. This also provides an understanding of the influence of each variable in the algorithm that shall be applied for the analytics. The next stage is the selection of the computing tools for the spatial analytics. Statistical programs are commonly used but using a Geographical Information Systems (GIS) is critical for spatial data analytics.

3.2 Spatial Analytics

Spatial Analytics is central to any data science project. Spatial analytics are targeted at different levels to transform data into information (Fig. 3).

The basic data analysis is descriptive analytics where the attributes and their trends are explored. An example could be to study the average day temperatures and see whether they are rising or falling over time for a region. Such findings are used for diagnostic analytics which is akin to a post-mortem of past events to find the prime influencing factors of an incident. E.g. deaths of people working on a hot and sunny day could be because of heat stroke. Once this is well understood, predictive analytics can be designed to flag trends which may lead to such an outcome and this helps is timely intervention to prevent the incident from happening in the future. E.g. One should not go to work under the sun when the temperatures are high. Prescriptive analytics is used where all the influencing factors can be identified to influence a desirable outcome. This is something like exploring ways to reduce the number of hot and sunny days in the future to reduce deaths from heat stroke.

Fig. 3 Levels of spatial data analytics

3.3 Algorithms

Each level of analytics is supported by one or more algorithms. An algorithm is a process or set of rules used in calculations by a computer. Spatial data science algorithms fall under 5 prime categories (Fig. 4).

Clustering divides the data into groups so that each group has similar properties. Several methods like hierarchical, density based, grid based, or fuzzy based algorithms are used to cluster datasets. Clustering helps in classification although classification is different as it uses pre-defined criteria. Decision trees, support vector machines (SVM), Bayes rule and neural networks are some examples based on how data can be clustered or classified. Regression establishes the numerical relation of an independent variable with a dependent variable. Regression methods can be linear, or polynomial depending on the complexity of the spatial distribution. Regression methods, in the past, have established previously unknown facts like the impact CO_2 concentration on temperature rise. But this does not necessarily establish the cause-effect relationship. Association rules establish this cause effect relationship where an antecedent is associated with a consequent.

Each class of algorithms are tweaked by researchers to suit the specific problem. Therefore, should an established algorithm yield unsuitable or biased results, it is essential to evaluate the sensitivity of the algorithm to obtain unbiased and logical results.

Fig. 4 Types of spatial data science algorithms

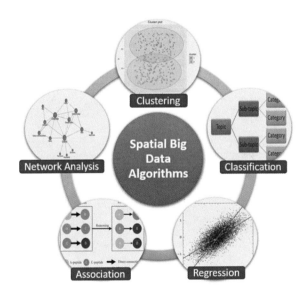

Fig. 5 Some areas of spatial data science applications in smart cities

4 Urban Applications

With the use of Information and Communication Technology (ICT) for operational efficiency in urban settings, spatial data analytics is being increasingly involved in advancing the Smart City concept. Singapore has been a leader in this aspect for using IOT applications for monitoring almost all activities. Other urban settings across the globe are following suit with plans to develop smart cities. This has been a driver of Spatial Data Analytics and applications where Big Data is utilized to provide insights into current activities. This helps in laying the plan for future activities involving people and commodities (Fig. 5).

4.1 Smart Sensors and Smart Cities

Smart sensors are being installed at critical locations where data is continuously collected. The application is spread across several services like energy management, traffic management, health and wellness, town planning and even in education.

Power usage patterns are gathered, and insights drawn to develop a smart grid where the production pattern preludes the consumption pattern thereby optimizing energy production and reducing energy wastage. Usage analytics are also shared to the consumer so that they can reduce usage and generate savings at their end. Smart signalling system is deployed to change traffic signals based on the number of waiting vehicles at road crossings. This reduces vehicular pollution from idling engines and also reduces travelling time of commuters. Network analysis helps in routing the shortest cost path for supply-chain management of consumer goods by reduceing wastage and lessening damage to the environment. Electronic health records pick up lifestyle related causes for health issues which when addressed impart a good quality of life. Doctors can also see patterns of people falling sick and can easily associate that with prevailing environmental factors like seasons or an environmental event. With the pandemic closing schools for longer periods, data science analytics are being used more to study the online educational activity patterns of pupils undergoing home-based learning. In the long run, these shall be used to evaluate learning objectives and outcomes and shall help the educator and the student alike.

4.2 Applications in This Book

In this book, each following chapter is a live demonstration of a spatial data science application in an urban setting.

Chapter 'Preferential Home Search in Urban Settings' is a study on people's considerations for searching a place to live in Singapore and demonstrates how spatial data can help them refine their search based on priorities. Major property portals have some spatial criteria that help home seekers, but this chapter identifies areas where things can be considerably improved. Chapter 'Air Quality Dynamics and Urban Heat Island effects during COVID-19' compares a COVID-19 setting against a pre-COVID-19 setting to understand air quality dynamic and urban heat island effects in Wuhan City and New York City. This provides an opportunity to identify similar patterns in the different geographies. Chapter 'Mass Rapid Transit and Population Dynamics during Covid-19 in Singapore' studies Mass Rapid Transit ridership in Singapore to identify weekly and weekend travel pattern amid COVID-19 lockdown and after some restrictions were lifted. Chapter 'A Geospatial analysis of Tweets during Post-Circuit Breaker in Singapore' demonstrates how spatial data analytics can be applied on twitter data for understanding relationships with service amenities and land use in urban setups. It is worth mentioning that tweets and other social media posts are now being increasingly used by governmental agencies for an indirect assessment of public sentiment on a myriad of issues. Chapter 'Space–Time Analytics of New York City Shooting Incidents' investigates the space-time analytics of New York Shooting which has an implication on allocation of policing administrative units to address the cases. Chapter 'Spatial and Temporal patterns of Tourist Source Market Emissiveness: A Study of Shanghai, China' deals on Tourism emissiveness in China with Shanghai being the focus of the study. Chapter 'Spatial Perspectives of Crime Patterns in Chicago amid Covid-19' studies the crime patterns in Chicago and reflects on how those patterns change with the onset of COVID related restrictions on anthropogenic activities. Chapter 'Geospatial analysis of Grab Trips in Singapore' evaluate the use of private car hires in Singapore in space and time based on data derived from the major hire car aggregator Grab. The final chapter, Chap. 'Ecological Vulnerability of Nyingchi, Tibet', is a study on spatial variability of ecological vulnerability in the city of Nyinchi in Tibet where insights from remote sensing data and Spatio-temporal analysis brings out the vulnerability zonation for addressing the severity of the problem.

Preferential Home Search in Urban Settings

Dina Labiba, Mark Frank Ratnam, Zhang Liqing, and Dini Aprilia Norvyani

Abstract Accessibility to amenities is an essential consideration for renting or purchasing a home. Therefore, ease of access and proximity to amenities is something Singapore's Housing and Development Board (HDB) emphasises on for the planning and development of housing towns in Singapore. Most amenities can be categorised as transport amenities, healthcare amenities or education amenities. Improving the availability and access to such amenities is an effective way to meet the housing needs of Singaporeans and HDB tries to ensure that the needs of Singaporeans are taken care of. This study presents a geospatial evaluation of access to various amenities from HDB blocks and housing towns. It features a proximity analysis (walking time and distance) to and from each amenity to HDB blocks. Factors like personal preference of individuals are considered in querying a suitable located HDB to assist home search using the Closest Facility Analysis tool in ArcGIS. Weights are assigned to distance to amenities to personalise home search and this provides a potential buyer with an additional tool to spatially query on a list of homes to choose from. The study shall provide potential buyers with a list of suitable options using which a multiple criteria search in terms of nearby amenities can be used to look for a suitable home. The study also aims to identify housing towns in Singapore which lack ease of accessibility to amenities in terms of time and distance and this can be used for planning of setting up amenities near these HDB blocks to make them more saleable. This study sheds light on housing towns which require improvements in the provision of amenities for residents.

1 Introduction

In high-density area like Singapore, compact and smart growth has always been a challenge and there have been significant gaps which require research and results put into practice. With a land area of only 714.1 km^2, Singapore has always facing challenges as to how land use can be optimised to meet the merging requirements

D. Labiba · M. Frank Ratnam · Z. Liqing · D. A. Norvyani (✉)
National University of Singapore, 21 Lower Kent Ridge Rd, Singapore 119077, Singapore

© The Author(s), under exclusive license to Springer Nature Singapore Pte Ltd. 2022   11
S. N. Kundu (ed.), *Geospatial Data Analytics and Urban Applications*,
Advances in 21st Century Human Settlements,
https://doi.org/10.1007/978-981-16-7649-9_2

growing population. It also aims to reduce the carbon footprint from travel by co-locating home, work locations and the amenities on needs to sustain life. Residents tend to choose to live in localities where everything they need is within walking reach.

1.1 HDB Towns in Singapore

To address the housing crisis in Singapore, the Housing Development Board (HDB) was set up in 1960 to address the problem within a span of 10 years [13]. In 1970s, a typical chessboard town planning structure emerged (Fig. 1), that constructed HDB towns based on the physical and spiritual needs of residents. Each HDB town consti-tuted a relatively independent mixed land use, with most daily amenities like child-care, eldercare, schools, clinics, and public transport stations available within the town. The main purpose of this chessboard structure [4] could be summarized as follows:

- integrate transportation and land use,
- optimize land use,
- achieve liveable urban life.

Today, there are more than 24 HDB towns built all over Singapore, whose planning, construction and management are still continuously evolving and maturing (Fig. 1).

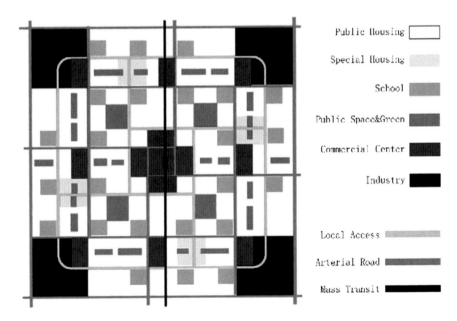

Fig. 1 Typical Singapore town structure [7] (*Source* HDB Singapore, 2020)

HDB housing, as the main part of the towns (Fig. 2), plays an important role to Singapore residents. According to the Annual Report 2018/2019 of HDB, more than 3.2 million of people (that is 80% of resident population) live in the HDB housing (Fig. 3). Thus, HDB towns are central to Singapore's housing needs and in effect to the well-being of Singaporean people.

Fig. 2 HDB towns in Singapore [6] (*Source* HDB Singapore, 2019)

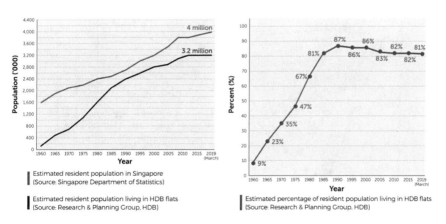

Fig. 3 Resident population in HDB housing [6] (*Source* HDB Singapore, 2019)

1.2 Amenities Within Walking Distance

The HDB town development comprehensively embodies the principles of public transport-oriented development and mixed use in walking distance. Western planning theories such as Neighbourhood Unit (Fig. 4), Traditional Neighbourhood Development (TND) and Transit-Oriented Development (TOD) (Fig. 5). These, in terms of location of amenities, transportation organization, and intensity development, occupy an important role in Singapore's current urban space development [2, 10, 12]. This compact development mode not only provides convenience to residents to execute their daily activities, but also provides commuters with seamless connections by public transport. This, in essence, reduces the reliance on the private cars, which is a progressive step towards sustainable urban development and practices.

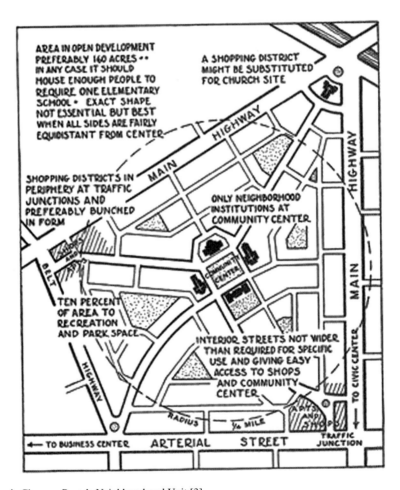

Fig. 4 Clarence Perry's Neighbourhood Unit [3]

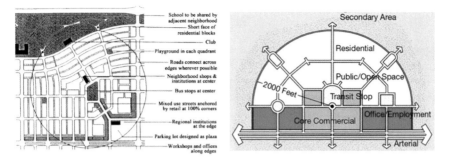

Fig. 5 Ilustration of TND (left) and TOD (right)

However, with the improvement of people's living standards, new requirements redefining residential quality have emerged. For sustainable development practices, people tend to adopt a healthy, safe and comfortable way of life. To support these emerging needs, the planning of new HDB town should be further improved to offer nearby daily amenities within walking or cycling distance. Singapore's Land Transport Master Plan 2040 [8] advocates a compact development mode called "20-min Towns", that means "commuters can expect to travel no more than 20 min via Walk-Cycle-Ride options for their daily amenities in their nearest neighbourhood centre and can access facilities such as parks and schools" (Fig. 6). Based on that, there will be demands when choosing HDB flats regarding the accessibility of nearby amenities.

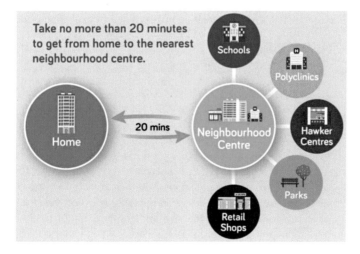

Fig. 6 20-min' walk-cycle-ride mode (*Source* LTA Singapore, 2020)

1.3   Searching for Property

Most in Singapore use online portals for their home search (e.g., www.propertyg uru.com.sg). Both property agents and direct sellers, advertise their property in these portals while buyers or tenants use these portals to search based on some geographical criterion. These web portal offers several search mechanisms like property type, location district, distance from Transportation node (Mass Rapid Transit Stations) etc. However, the spatial search filters do not provide any proximity analysis to many other amenities like schools, hospitals etc. A potential buyer or tenant, therefore, has to be geographically aware through other maps tools which are not integrated in the property portal. They may also have to take the pain to travel to the locality to physically checkout on the nearby amenities.

The erstwhile Housing and Development Board (HDB) CEO, once discussed how the HDB has planned and designed HDB housing towns through the use of technology and community outreach to develop better homes [5]. It was mentioned that in Singapore, HDB flats are 'like the air we breathe'. With more than 80% of Singapore's population living in HDB housing, there is a high chance that someone would have made use of some of the amenities found in an HDB town, such as the market or hawker centre, the neighbourhood shop or clinic. These amenities should be located near to housing towns for convenience and public access.

Thus, accessibility and proximity to amenities such as childcare, clinics, eldercare facilities and transport amenities are critical considerations for potential buyers and tenants. Therefore, a spatial search tool to filter and compare housing regarding the accessibility of amenities visually in property portals would be invaluable greatly add value to the users. This study aims to provide insights to fill this gap by proposing an analysis on how such amenities can be incorporated in one's home search.

2   Study Objectives

2.1   Objectives

Based on [9] around 82% of all Singaporean are living in a flat house that provided by Housing Development Board (HDB). There are total of 24 HDB townships in Singapore. Each HDB are unique in terms of their proximity to various amenities as they we built in different time under different planning objectives. Preferences of families is natural to vary based on the size and composition of the household. A family with is likely to prefer HDBs that is more accessible to education resources as compared to it being in vicinity of a transport node. Singles, on the other hand, prefer to live near public transport facilities rather than education institutions. Therefore, in this study we attempt to create a simple tool that could assist a buyer or tenant in choosing their preferred HDB areas amongst the many choices available. This study also shall provide information on HDB housing performance through a proximity analysis to preferred amenities.

One prime objective of this research is to create additional search criteria to improve the search experience of buyers and tenants in online property search portals by including search criteria based on proximity to amenities such as schools, clinics, childcare, eldercare, supermarkets in addition to MRT stations. With this potential buyers and tenants should be able to streamline their search according to their needs and priorities. Including multiple proximity criteria based on individual preference weights is also considered in this study. With rich spatial data being publicly available, more amenities can also be later added to the search criteria in the future too. Moreover, with data on more amenities, one would be able to visualize the distribution of HDB areas by rank based on people preferences.

Another objective of this research is also to identify HDB towns which are lacking in terms of accessibility to the amenities. A methodology of ranking HDBs based on available assessable amenities in proximity is discussed which can provide insight on future HDB town planning. Suggestions for improving accessibility to amenities for HDB towns shall be discussed and a dashboard shall be demonstrated on how the customers can choose an HDB based on the distance to amenities.

2.2 Scope

This study includes all the HDB blocks and HDB towns in Singapore. The proximity analysis is planned based on the 20-min walking distance to certain amenities which include Childcare, Eldercare, Schools, Clinics, Bus stops and MRT stations.

3 Data and Methodology

3.1 Data Types and Sources

The data and sources that were used in this research are listed in Table 1. The data used in this research was primarily obtained through data sources from data.gov.sg and MOE-EDUGIS (a GIS repository of data that has been mined and collated for use in schools in geospatial applications and projects).

3.1.1 HDB Location

The data for HDB buildings was obtained from the MOE-EduGIS portal. The data was curated by the Geography Unit, Ministry of Education (MOE) from the Housing and Development Board's Geospace files: HDB Building Age and HDB existing

Table 1 Data and Sources

Data	Description	Source
HDBs location	Location data of HDB blocks in Singapore year 2018 Spatial data (.shp) Geometry Type: Point	Geospace/MOE EduGIS
Public transport amenities	Location data of the hub/transit of Singapore road and rail transports, including buses, and MRT station exits Spatial data (.shp) Geometry Type: Point	Geospace/MOE EduGIS
Health amenities	Location data of CHAS clinics Spatial data (.shp) Geometry Type: Point	data.gov.sg
Care amenities	Location data of childcare and eldercare Spatial data (.shp) Geometry Type: Point	data.gov.sg
Education amenities	Location data of schools Spatial data (.shp) Geometry Type: Point	MOE EduGIS

building. The map data layer contains polygons outlining the locations of HDB buildings as well as information such as postal code, year of completion, purpose (residential/commercial), presence of food centre or multi-storey carparks. The data was based on 2018 information and contained 11,924 data entries. The original polygon data was then converted into point data for analysis.

3.1.2 Public Transport Amenities

The data for public transport amenities such as bus stops and MRT station exits were also obtained from the MOE-EduGIS portal. The data for bus stops was obtained from the Land Transport Authority's Geospace file: LTA Bus Stop. The data was based on 2019 information and contained 4882 data entries. The data for MRT station exits was obtained from the Land Transport Authority's Geospace file: LTA MRT Station Exits. The data was based on 2019 information and contains 468 data entries.

3.1.3 Health and Care Amenities

The data for health amenities was obtained from the Singapore Government data portal (www.data.gov.sg). The data provides the location of CHAS (Community Health Assist Scheme) Clinics in Singapore managed by the Ministry of Health. The data was last updated in 2020 and contains 1163 data entries.

The data for care amenities was obtained from the Singapore Government data portal (www.data.gov.sg). The data for childcare facilities was provided by the Early Childhood Development Agency based on updated information from August 2020. The childcare data in KML format contains 1545 data entries. The data for eldercare facilities was obtained from the Ministry of Social and Family Development based on information from April 2017. The eldercare data contains 133 data entries.

3.1.4 Educational Amenities

The data for schools was also obtained from the MOE-EduGIS portal. The data provides the locations of different categories of schools ranging from Primary, Secondary, Mixed levels and Pre-University schools. The data from the Ministry of Education is based on information from 2020 and contains 344 data entries.

3.2 Methodology

For this study the prime tasks are listed as below.

- Data Source Identification
- Data wrangling
- Database Management
- Algorithms Selection and Application
- Analysis and evaluation of results.

The workflow (Fig. 7) summarizes the next level of details employed for the study. The methodologies which were applied on spatial data are categorized into two.

- Proximity Analysis
- Simulation and Visualization

3.2.1 Proximity Analysis—Closest Facility

Proximity analysis of closest facility was done to calculate the distance of each HDB block to each of the nearest amenities. Closest facility was used to model and calculate the shortest distance along the street network from each point incident to any facilities [11, 14]. The general equation that refers to previous study (e.g., [14] is defined as below:

$$A_i^t = Oj \tag{1}$$

$$\sum A_i^t = \sum_{j \in Ni} Oj \tag{2}$$

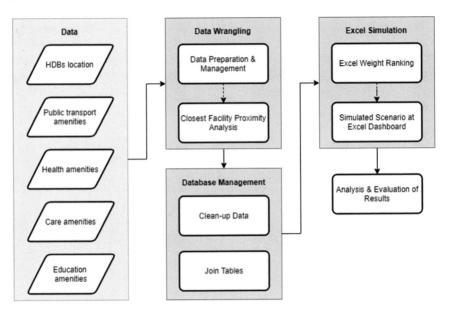

Fig. 7 Project workflow

where Ati is the distance of amenities within HDB town i with a given travel threshold of t, Oj equals one and represents a distance route to location j, and in this study, we only consider the first rank facility that will be included in the calculation, and Ni is the subset of amenities [14]. Referring to our proposed study objective, which is to analyse the shortest distance based on walking-travel mode, the travel threshold will be in distance (meters). In this study, HDB block will be perform as the incident point and each amenity will be the facility point.

For the assigned amenities, we are referring to the amenity theory that shaped the city. There are two different types of amenities within neighbourhood, exogenous and endogenous. Exogenous amenity is generated by the history and natural of the city, while endogenous are generated by the socio-economic condition of the resident within the city [1, 9]. In this study, we are more focusing on the endogenous type of amenity that influence Singapore people in choosing their living town preference. List of amenities that we are using are childcare, eldercare, bus stop, MRT station, school and clinics that should be accessible from each HDB block. The result of the analysis will be stored in the form of a database table that will be further used for generating the weightage tools.

The distance results from closest-facility analysis then will be managed and stored as a database for analysis and processing in the information system. The distance database will be differentiated into two types: (a) closest facility routes of each HDB blocks; and (b) closest facility routes of HDB town; This result will be used to evaluate HDB towns performance in terms of distance to amenities.

3.2.2 Simulation and Visualization

A dashboard was created, together with the algorithm/formulation to enable users to input their preference criteria (weightage) for amenities that they consider important to them. The input from users will be processed in the formula:

$$D_w = D_{Cc}W_{Cc} + D_{Ec}W_{Ec} + D_c W_c + D_S W_S + D_B W_B + D_{MRT}W_{MRT} \qquad (3)$$

A constraint is applied for this formula, as follows:

$$W_{Cc} + W_{Ec} + W_c + W_B + W_{MRT} = 1 \qquad (4)$$

where Dw is weighted distance; DCc, DEc, DC, DS, DBS, and DMRT respectively are distance from HDB to childcare, eldercare, clinics, school, bus stop, and MRT. Where W is weight for each amenity's accessibility. Finally, based on the computation from the formula, the dashboard will give the recommendation of the most suitable HDBs based on their input.

4 Results

4.1 Closest Facility Analysis

Most of the distance from HDB blocks to the amenities, as the result of closest facility are within 1,600 m, or equally can be reached within a 20-min-walk. Childcare and bus stops are the most reachable amenities within each HDB towns where most people can reach it within 600 distance walks; While Eldercare is the most distant facility to reach by residents where the range to reach the eldercare facility is within 600–3,700 m (Fig. 8).

Figure 8 shows that each HDB has their own distance characteristics to certain amenities. For example, referring to the average distance to childcare (Fig. 9a), most HDB block in Bukit Timah town can reach any childcare within 62 m, which can be assumed as the most accessible HDB town in terms of childcare. However, the average distance to reach any other assigned amenities (bus stop, clinics, MRT station) are one of the furthest compared to other HDB towns. In Sembawang (Table 2), for example, the average distance to reach Eldercare facility is within 3,447.4 m, one of towns with the farthest distance to reach the Eldercare facility compared to others. However, when we look at the average distance to reach Bus Stop, Sembawang HDB town can reach it within 86.4 m (less than 5 min-walk), the shortest distance compared to others.

Based on the result, we can see that each HDB has their own characteristics in terms of accessibility to amenities. According to that, people will have their own preference in choosing their favourable HDB towns in terms of amenities. People

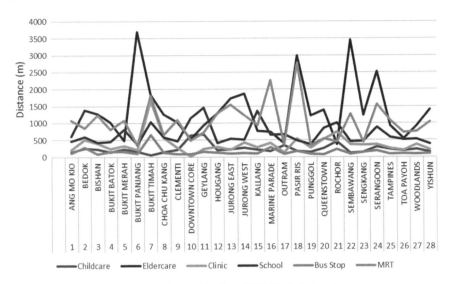

Fig. 8 Graph of average distance of amenities from HDB blocks in housing towns

that are living with elderly or for elder group people, they might prefer to live in HDB town which has the shortest distance to eldercare, while a group of family that having school-aged children, might prefer to live in HDB town which has the shortest distance to school.

4.2 Improvement Analysis

According to Fig. 8, the average proximity of various amenities within each housing town varies. This could be due to contributing factors such as population demographics, year of establishment of housing units as well as amenities. The range of completion dates of HDB blocks within housing towns range from 1937 to 2017 (as of the last year recorded in the data). HDB housing towns are constantly evolving and changing. Neighbourhood renewal is one of those programmes implemented by HDB that results in constant changes to the layout and structure of housing towns in Singapore. This perhaps was the reason it is useful to study the accessibility of housing towns in relation to amenities in line with neighbourhood renewal programmes.

4.2.1 Eldercare

Based on Fig. 9, several housing towns are noticeably lacking in proximity to certain amenities. These towns are Bukit Panjang, Pasir Ris, Sembawang and Serangoon. Housing towns that require some improvement in their provision of amenities include

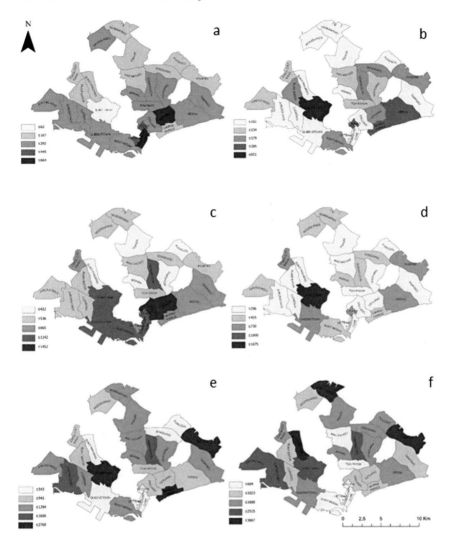

Fig. 9 Singapore average distance (in meters) to each amenity mapping using closest distance analysis—**a** distance to childcare; **b** distance to bus stop; **c** distance to school; **d** distance to clinics; **e** distance to MRT; **f** distance to eldercare

Bukit Panjang (eldercare), Bukit Timah (eldercare and MRT), Pasir Ris (eldercare and MRT), Sembawang (eldercare) and Serangoon (eldercare).

Thus, one of the most common areas for improvement in housing towns are required in the provision of eldercare facilities. In a country like Singapore which faces an ageing population, it is important for the relevant government agencies to look into providing sufficient eldercare facilities in the housing towns. It is noted that towns like Pasir Ris are relatively matured and therefore lack the necessary

Table 2 List of average distance of amenities from HDB blocks in housing towns

HDB towns		Childcare (m)	Eldercare (m)	CHAS Clinic (m)	School (m)	Bus Stop (m)	MRT (m)
1	ANG MO KIO	145.3	604.0	212.7	470.2	117.0	1,075.2
2	BEDOK	260.2	1,390.8	508.5	593.9	282.8	855.7
3	BISHAN	228.0	1,272.5	403.1	422.4	87.6	1,230.0
4	BUKIT BATOK	145.8	1,014.7	241.2	453.7	160.1	801.1
5	BUKIT MERAH	232.2	483.1	327.9	809.9	161.3	1,082.7
6	BUKIT PANJANG	160.1	3,687.5	219.4	375.4	95.5	342.3
7	BUKIT TIMAH	62.0	1,835.3	1,675.4	1,032.4	650.6	1,807.6
8	CHOA CHU KANG	147.0	1,255.5	571.8	584.5	133.7	653.3
9	CLEMENTI	217.2	1,022.5	269.3	474.3	92.8	1,096.1
10	DOWNTOWN CORE	632.1	530.1	21.7	1,142.4	86.2	486.5
11	GEYLANG	664.4	928.7	248.4	1,452.1	162.2	725.3
12	HOUGANG	204.3	1,286.3	295.8	405.5	126.2	1,294.0
13	JURONG EAST	221.8	1,722.0	215.7	536.0	95.7	1,550.1
14	JURONG WEST	263.8	1,872.8	425.4	508.7	108.1	1,232.7
15	KALLANG	212.6	765.2	276.9	1,355.2	99.3	941.2
16	MARINE PARADE	166.5	736.5	409.2	655.8	285.2	2,260.9
17	OUTRAM	335.8	396.9	128.0	664.6	86.4	361.0
18	PASIR RIS	166.5	2,993.0	537.5	484.4	165.0	2,768.8
19	PUNGGOL	138.8	1,218.1	263.9	374.0	86.3	340.9
20	QUEENSTOWN	260.4	1,391.6	518.0	850.5	70.1	543.5
21	ROCHOR	448.6	362.5	730.1	1,002.9	222.2	459.7
22	SEMBAWANG	117.2	3,447.4	369.2	453.5	86.4	1,272.0
23	SENGKANG	140.6	1,231.5	353.5	467.8	149.0	446.0
24	SERANGOON	292.1	2,515.3	362.0	884.7	179.4	1,557.1
25	TAMPINES	221.3	923.3	262.4	576.1	85.8	1,082.5
26	TOA PAYOH	191.5	538.0	199.3	507.5	105.2	729.5
27	WOODLANDS	233.3	917.5	363.4	518.9	88.4	757.4
28	YISHUN	141.7	1,397.4	216.0	383.2	101.3	1,037.2

infrastructure for eldercare facilities. Thus, conscious plans are required to convert existing amenities into more desired amenities such as eldercare. As the population in Singapore becomes more aged, families are increasingly responsible for taking care of their elderly parents. Therefore, being close to eldercare facilities would be an increasing priority to families looking to purchase their homes.

4.2.2 MRT Station

Another noticeable amenity that requires improvement is the availability of MRT station exits close to HDB blocks. Table 2 provides information on the housing towns which possess relatively higher average distances to MRT stations. These housing towns are Bukit Timah, Marine Parade and Pasir Ris. The average distance of MRT stations to HDB blocks in Bukit Timah is 1,807.6 m, Marine Parade is 2,260.9 m and Pasir Ris is 2,768.8 m. According to the LTA Masterplan, it is targeted that every HDB block will be within a 10-min walk to an MRT station. This is perhaps why this target has not yet been achieved and the LTA is working hard with SMRT and SBS Transit to increase the rail network by increasing the number of MRT lines. Thus, the availability of MRT stations near HDB blocks will be something that will be improved upon in the future years.

4.2.3 Bus Stops

While the average distance of bus-stops from HDB blocks are generally below 650 m, there are variations in average distances of bus-stops between housing towns. The average distance to bus stops from HDB blocks within towns such as Bedok and Marine Parade are greater than 285.2 m while those within Bukit Timah are greater than 650.6 m. It could be argued that HDB blocks may not be as close to bus stops as compared to other housing towns due to the socio-economic status of residents living within Bukit Timah. This is because Bukit Timah is mainly populated by upper/middle class residents who may not depend as much on buses for transportation as they are more likely to take private transport such as cars or taxis.

One of the contributing reasons why Bedok may have such poor accessibility to bus stops could be due to the fact that Bedok is a matured housing town and most of their HDB blocks and road networks have already been set in place. There is thus a need to relook into the bus-routing networks and bus stops within Bedok town so that the average distance of bus stops from HDB blocks can be reduced to meet the commuting needs of residents.

4.2.4 Schools

The majority of housing towns in Singapore are within 1,000 m on average between schools and HDB blocks. Thus, it would seem that the provision of educational amenities is sufficient and meet the needs of residents. Table 2 shows that there are several housing towns which exceed 1,000 m in terms of average distance to schools. These towns are Geylang (1,452.1 m), Kallang (1,355.2 m), Downtown Core (1,142.4 m), Bukit Timah (1,032.4 m) and Rochor (1,002.9 m). However, although these towns exceed 1,000 m, they are still within the accepted 1,600-m radius for amenities as stipulated in this research earlier. Another noteworthy point is that these towns which are above 1,000 m are not the usual HDB housing towns that the majority of Singaporeans reside in. That could possibly explain the lack of closeness to educational schools in these towns.

4.2.5   CHAS Clinics

Table 2 shows that with regards to clinic amenities, the only housing town which does not meet the stipulated maximum distance of 1,600 m for amenities is Bukit Timah. The average distance of CHAS clinics in Bukit Timah is 1,675.4 m. A possible explanation for this could be the fact that CHAS clinics are primarily catered to provide subsidized healthcare to residents. The higher socio-economic status of residents in Bukit Timah could be the contributing factor why there are fewer CHAS clinics within proximity to HDB blocks within Bukit Timah. In fact, there would be a greater concentration of private and specialist clinics within the Bukit Timah area rather than CHAS clinics. In order to provide a more equitable distribution of subsidized healthcare to all Singaporeans, there should be an increase in the provision of CHAS certified clinics within Bukit Timah housing town.

4.2.6   Childcare

While the average distance of childcare amenities from HDB blocks are generally below 665 m, there are variations in average distances of childcare amenities between housing towns. Figure 9a shows that Geylang and Downtown Core are the two housing estates with the greatest average distances between childcare amenities and HDB blocks. The average distance to childcare from HDB blocks within Geylang is 664.4 m and Downtown Core is 632.1 m, according to Table 2. Although more could be done to improve the accessibility of childcare amenities in Geylang and Downtown Core, it is also necessary to note that Geylang is a rather matured estate with an ageing population, thus having a smaller percentage of children residing in Geylang. This could explain the lack of childcare amenities within Geylang. Downtown Core on the other hand is a popular housing location for young and trendy newly married couples who have relatively successful careers and wealth. Thus, there is a need for the authorities to consider providing more childcare amenities within the Downtown Core area as these younger couples may be starting their families in due time.

5 Discussion

5.1 Simulation Dashboard

An excel-based simulation dashboard was developed as a part of this study. This tool supported simulation of individual's custom preferences for choosing a property based on proximity to the six considered amenities. Each preference could be weighted based on their priorities (Fig. 10).

Based on the weighted amenity distance of HDB blocks and towns, simulations were done by considering the percentage of population age group. Therefore, there are three scenarios applied: children, productive, and elderly. The simulations examine whether the recommendation given based on the inputted weight relevant to the demography of the HDB towns. The applied weight for each scenario is shown in Table 3.

The dashboard suggests the most suitable HDB towns and/or HDB blocks based on the weighted inputs. Changing the weights based on personal preference for each scenario, produces different recommendations for customer in choosing the HDB and can be visualized in a map Fig. 11.

Using the assumption that household with children will prefer to have proximity to schools, 0.25 weight was applied to the distance to school amenities, for children age group scenario. Excel simulation for this age group scenario recommends Punggol,

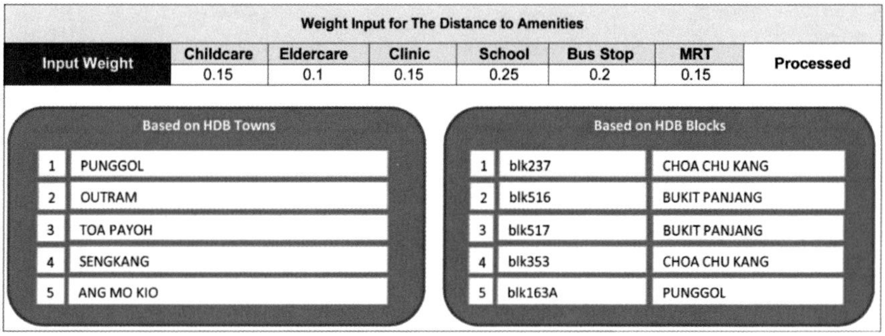

	Weight Input for The Distance to Amenities						
Input Weight	**Childcare**	**Eldercare**	**Clinic**	**School**	**Bus Stop**	**MRT**	**Processed**
	0.15	0.1	0.15	0.25	0.2	0.15	

	Based on HDB Towns			Based on HDB Blocks	
1	PUNGGOL		1	blk237	CHOA CHU KANG
2	OUTRAM		2	blk516	BUKIT PANJANG
3	TOA PAYOH		3	blk517	BUKIT PANJANG
4	SENGKANG		4	blk353	CHOA CHU KANG
5	ANG MO KIO		5	blk163A	PUNGGOL

Fig. 10 The Simulation Dashboard

Table 3 Applied weight for each age group-based scenario for choosing the HDB based on distance to amenities

Age Group	Applied weight for the distance to Amenities					
	Childcare	Eldercare	Clinics	School	Bus Stop	MRT
Children	0.15	0.10	0.15	0.25	0.20	0.15
Productive	0.15	0.10	0.10	0.15	0.25	0.25
Elderly	0.10	0.25	0.20	0.10	0.20	0.15

Fig. 11 Visualization of
HDB towns ranking based
on scenario simulation—**a**
from children age-group; **b**
from productive age-group; **c**
from elderly age-group

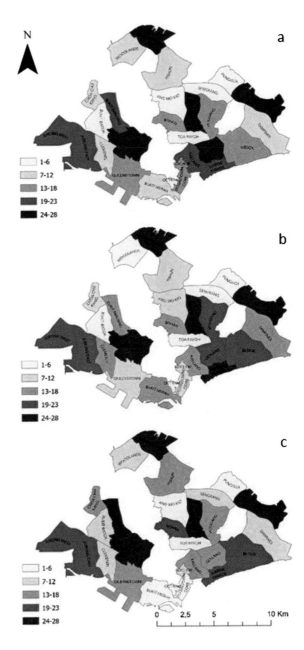

Outram, Toa Payoh, Sengkang, and Ang Mo Kio for the customers. Based on the demography profile, two towns match this recommendation, which are Punggol and Sengkang. For the productive age group, the biggest weight was set to the distance to bus stop and MRT, considering the high mobility for the people who commute to work. Last, for the elderly age group, biggest weight was set to the distance to eldercare, followed by clinics.

However, referring to Fig. 11, all towns with least average distance to each amenity are not necessarily being recommended. This is because the variable considered is not just for a particular amenity and the distance to other amenities contribute to the search despite their weights being dissimilar. Also, it applies to a particular HDB blocks, and not to the average of all blocks in each HDB town (Table 4).

5.2 Limitations

Based on the findings, we found that this research has several limitations, as follow:

Inconsistent year of data collected.

Due to the nature of the data being mined from various agencies and sources, it is difficult to accurately obtain data from the same point in time for accurate analysis and comparison. Thus, while some data may be older than others, the resulting relationships and outcomes may have changed over time.

Dynamic changing in HDB towns.

Housing towns are constantly being renewed and rebuilt. Older buildings and facilities are being torn down and new facilities and housing blocks are being built in existing housing towns. Therefore, the changing dynamics of population, population needs, and the built environment present a challenge in being able to accurately present a comprehensive analysis of the housing town as it is.

Limitation in Singapore Road Network Dataset.

Due to the data limitation in Singapore network dataset, where the available data is not fully complete and there is no Singapore pedestrian network, there are some amenities and HDB block that cannot be identified by the network dataset. Thus, some HDB block may not be included in the analysis or not have the closest distance value. Future analysis might be needed with using the more complete network dataset to result more accurate closest distance analysis.

Table 4 Top 5 recommendations for 3 scenarios

Age group	5 most recommended HDBs				HDB towns with highest percentage of age group population	
	HDB towns		HDB blocks			
Children	1	PUNGGOL	1	Blk237, CHOA CHU KANG	1	PUNGGOL
	2	OUTRAM	2	Blk516, BUKIT PANJANG	2	SENGKANG
	3	TOA PAYOH	3	Blk517, BUKIT PANJANG	3	SEMBAWANG
	4	SENGKANG	4	Blk353, CHOA CHU KANG	4	BUKIT TIMAH
	5	ANG MO KIO	5	Blk163A, PUNGGOL	5	WOODLANDS
Productive	1	OUTRAM	1	Blk237, CHOA CHU KANG	1	PASIR RIS
	2	PUNGGOL	2	Blk516, BUKIT PANJANG	2	CHOA CHU KANG
	3	TOA PAYOH	3	Blk105, KALLANG	3	WOODLANDS
	4	SENGKANG	4	Blk517, BUKIT PANJANG	4	TAMPINES
	5	WOODLANDS	5	Blk242, BUKIT PANJANG	5	BUKIT BATOK
Elderly	1	OUTRAM	1	Blk237, CHOA CHU KANG	1	OUTRAM
	2	TOA PAYOH	2	Blk302, GEYLANG	2	ROCHOR
	3	DOWNTOWN CORE	3	Blk201A, PUNGGOL	3	BUKIT MERAH
	4	ANG MO KIO	4	Blk516, BUKIT PANJANG	4	TOA PAYOH
	5	PUNGGOL	5	Blk353, CHOA CHU KANG	5	QUEENSTOWN

5.3 Conclusions

Amenities with high accessibility within walking and cycling distance to the residence comply with the sustainable and compact development, especially in the high-density country like Singapore. HDB housing, as the most common residence form in Singapore, the accessibility of daily amenities around it is closely linked with residents' well-beings. This research concludes that:

- Most prevalent online flat form for flats searching, has certain limitation in filtering and comparing flats in terms of amenities' accessibility. A dashboard in excel can help buyers to filter out their ideal flats by inputting personalized weight to each amenity.

- Six rankings choropleth maps of all HDB towns based on the averaged distance of each blocks to six types of amenities shows the towns with desirable performance as well as those need further improvement.
- Simulations based on three typical age groups show the recommended HDB towns and/or blocks depends on weight combinations of all amenities.

References

1. Brueckner JK, Thisse J-F, Zenou Y (1999) Why is central Paris rich and downtown Detroit poor? An amenity-based theory. Eur Econ Rev 43(1):91–107
2. Calthorpe P (1993) The next American metropolis: ecology, community, and the American dream. Princeton Architectural Press
3. Chaira J, Panero J, Zelnik M (1984). Time saver standards for housing and residential development. McGraw hill, Inc
4. Dai D, Yao D (2013) The research on the hierarchy of spatial structure's change and adaptability of Singapore. Planners S2:70–73
5. Hean CK (2018) Shaping the future of Singapore's heartlands. https://www.todayonline.com/commentary/shaping-future-singapores-heartlands
6. Housing and Development Board of Singapore (2019) HDB annual report 2018/2019: key statistics (pp 13–14). Housing and Development Board of Singapore. https://services2.hdb.gov.sg/ebook/AR2019-keystats/html5/index.html?&locale=ENG&pn=13
7. Housing and Development Board of Singapore (2020) HDB history and towns—housing & development board (HDB). Housing & Development Board (HDB). https://www.hdb.gov.sg/cs/infoweb/about-us/history
8. LTA (2020) Land transport master plan 2040. Land transport authority of Singapore. https://www.lta.gov.sg/content/dam/ltagov/who_we_are/our_work/land_transport_master_plan_2040/pdf/LTA%20LTMP%202040%20eReport.pdf
9. Lee J, Irwin N, Irwin E, Miller HJ (2020) The role of distance-dependent versus localized Amenities in polarizing Urban spatial structure: a spatio-temporal analysis of residential location value in columbus, Ohio, 2000–2015. Geogr Anal *n/a*(n/a). https://doi.org/10.1111/gean.12238
10. National League of Cities (2020) Traditional Neighborhood Development. National League of Cities. https://www.nlc.org/resource/traditional-neighborhood-development/
11. Nicoară P-S, Haidu I (2014) A GIS based network analysis for the identification of shortest route access to emergency medical facilities. Geogr Tech 09(2):60–67
12. Qian L (2006) From the neighborhood unit to the community of new urbanism: research on the evolution of the community planning model in USA. World Arch 7:92–94
13. Scheer B (2004) Suburban form: an international perspective. Psychology Press
14. Widener MJ (2017) Comparing measures of accessibility to urban supermarkets for transit and auto users. Prof Geogr 69(3):362–371. https://doi.org/10.1080/00330124.2016.1237293

Air Quality Dynamics and Urban Heat Island Effects During COVID-19

Liu Weiyu, Xu Yuanyuan, Sun Tong, and Wang Jifei

Abstract The outbreak of novel coronavirus pneumonia was the most serious global issue in 2020, that caused enormous impacts on various aspects of human society from public health to economic well-being. Our ecological environment also experienced transformation due to restricted human activities during the epidemic. The implementation of 'Lockdowns' and 'Stay-at-Home' policies reduced the pollution emissions from transport and commercial activities altering the urban environment. In this study, the spatial and temporal distribution and changes of the air quality and thermal environment in Wuhan City and New York City in 2020 were conducted and analyzed. Spatial data from these two cities were acquired, processed, interpolated and analyzed to identify hot spots and cold spots which were reasoned. The dynamics of air pollution and Urban Heat Island (UHI) effect was the prime object of investigation. The study discovered interesting patterns of changes in those two cities which we the early epicenters of the pandemic.

1 Introduction

The novel coronavirus pneumonia (COVID-19), which broke out in Wuhan in January and was characterized as a global pandemic on March 11, spreads rapidly during the past nine months and significantly impacted public health, economy, environment, and social interaction. COVID-19 became a long-standing challenge facing humanity and shall continue to be so for times to come. Several scientific studies are currently being done which focus on the geospatial and spatial-statistical analysis of COVID-19 pandemic. These include Spatio-temporal mapping of the disease spread and used method like social and environmental geography, data mining and web-based geo-analysis [6].

Among past work which investigated spatial autocorrelation of COVID-19, the Moran's Index was primarily used to map the distribution of epidemic infection cases. Researchers made predictions via logistic model in the early study [7, 8]. The spatial

L. Weiyu · X. Yuanyuan · S. Tong · W. Jifei (✉)
National University of Singapore, 1 Arts Link, Kent Ridge 117570, Singapore

© The Author(s), under exclusive license to Springer Nature Singapore Pte Ltd. 2022
S. N. Kundu (ed.), *Geospatial Data Analytics and Urban Applications*,
Advances in 21st Century Human Settlements,
https://doi.org/10.1007/978-981-16-7649-9_3

features of disease transmission were explored using Kernel Density Analysis and Ordinary Least Square regression on social media data as well, [14]. More factors were taken into consideration to understand the spatial patterns of COVID-19 using cluster analysis to evaluate the contextual factors [5].

Air quality is an indicator in environmental condition and is predominantly used for climate change study. According to Daniella et al. [15], air quality has improved across 50 cities during the lockdown period. The study conducted a global analysis of the relationship of air quality and population correlating PM2.5 concentration data and its variation before and after quarantine. Data from air quality stations in Brazil was analysed to assess air pollutant concentration variations during the partial lockdown too [12].

Urban Heat Island (UHI) is a micro-climate phenomenon in which urban areas have higher surface temperatures than the surrounding rural areas [23]. The main cause of UHI effect is the modification of urban configuration due to accelerating urbanization and the increasing anthropogenic activities. The magnitudes of the UHI effect are closely correlated with socio-economic activity and population increase. In most UHI studies, two kinds of data are widely applied in the assessment and quantification of the UHI effect. Traditionally, air temperature data from in-situ stations have the limitation of spatial coverage. But with the development of remote sensing techniques, land surface temperature (LST) derived from satellite images had been predominantly used for surface UHI (SUHI) studies. LST data is highly available and is can be of high spatial resolution [20]. Urban thermal characteristic is a significant component of the urban environment monitoring and during COVID-19, the restriction of human activities helped in the global restoration of climate elements such as air quality and temperature. Most recent articles have suggested that the air quality during the pandemic lockdown had improvement at a local or global scale, however, limited research has been done on urban thermal dynamics change in this period. LST, therefore, can be used to further evaluate the impact of COVID-19 on urban environment [11].

Most research during COVID-19 focused on the prediction and spatial distribution of the disease. A select few conducted statistical analysis of the socio-ecological and environmental consequences of the epidemic. The need of the hour is to conduct interdisciplinary analysis of impacts caused by COVID-19 as this shall be the basis to evaluate and update existing policies of health interventions and controls and for formulating plans for social life-style rehabilitation, economic recovery and environmental protection.

1.1 Objectives

To investigate the consequences of epidemic control strategies and understand the environmental changes during the disease outbreak, this research was conducted on two typical cities, Wuhan City and New York City, which were the early epicentres of the pandemic. The primary purpose of this study was to conduct a Spatio-temporal change analysis of air pollution (i.e. PM2.5) and thermal environment (i.e. land surface temperature) in the metropolitan urban area of these two cities. Using remote sensing data and geospatial analysis tools, the consequences of COVID-19 based on the distribution characteristics of disease mapping was further explored.

In a nutshell, the prime objectives of this study were as below are summarized as follows:

1. To understand the temporal and spatial distribution of COVID-19 cases in the two cities during the outbreak.
2. To assess the influence of epidemic spread based on the air quality level using critical air quality indicators, e.g. PM2.5, NO_2, CO, SO_2, O_3.
3. To investigate the spatial patterns of air pollutants and understand the Urban Heat Island effect before and during the spread of COVID-19.

1.2 Study Area

The study area are the two early epicenters of Covid-19 pandemic; Wuhan City and New York.

Wuhan City which is located in Hubei Province, China (Fig. 1 left), was where the disease first originated. The Yangtze River flows through the center of Wuhan city, dividing the central urban area of Wuhan into three parts. The city has a population of over 11 million and underwent a lockdown from January 23 to April 8. Until September 8, 2000, there were a total of 68,139 cases of Covid-19 infections [14].

New York City (Fig. 1 right). which is located at the southern tip of the US, became the next epicentre of the epidemic towards the end of March 2020. The city has a population of 8.3 million people spread across five boroughs interconnected by

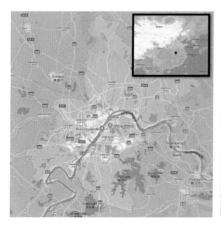

Fig. 1 The study areas: Wuhan, China (left) and New York, USA (right)

a subway system that also extends into neighboring areas [17]. The city had 445,071 cases as of September 8, 2020.

2   Data and Methods

2.1   Data

For the comprehensive analysis of various environmental factors, data was sourced from multiple agencies (Tables 1 and 2). They can be categorised as below.

- The COVID-19 cases, population composition, and hospital assistance data.
- The daily air quality data of the Wuhan City and New York City.
- Satellite imagery data covers the study area.

Table 1 Data types and sources

Data	Wuhan	Website
COVID-19	Wuhan Municipal Health Commission	http://wjw.wuhan.gov.cn/xwzx_28/
Air quality data	China National Environmental Monitoring Centre	http://www.cnemc.cn/
Satellite imagery	USGS	https://glovis.usgs.gov/app
Data	New York City	Website
COVID-19	NYC Health	https://www1.nyc.gov/site/doh/covid/covid-19-data.page
Air quality data	United Stated Environmental Protection Agency (EPA)	https://www.epa.gov/outdoor-air-quality-data
Satellite imagery	USGS	https://glovis.usgs.gov/app

Table 2 Landsat 8 OLI/TIRS images data acquisition dates

Location	Pre-Event	Post Event
Wuhan China	2019/03/10	2020/02/09
	2019/06/14	2020/04/13
	2019/08/01	2020/08/03
	2019/10/20	2020/10/22
New York City U.S	2019/02/19	2020/03/09
	2019/04/24	2020/05/12
	2019/05/26	2020/06/13
	2019/06/27	2020/10/03

The COVID-19 data for Wuhan was sourced from the Wuhan Health Commission website. The equivalent data for New York City was sourced from NYC Health website. COVID-19 cases data in Wuhan between 23rd January to 28th May was used whereas for New York City data from 1st March to 17th June was used. The daily Air Quality Index (AQI) data of Wuhan, from 1 January 2016 to 31 September 2020, was provided by the China National Environmental Monitoring Centre. The daily AQI data of New York, from 1 January 2016 to 31 September 2020, sourced from the United States Environmental Protection Agency (EPA).

The Landsat 8 satellite images for the year 2019, 2020 (Table 2) were sourced from United States Geological Survey (USGS). Imageries with little cloud cover over the study area were chosen, one before and one after the pandemic outbreak. The Landsat 8 satellite sensors comprised the Operational Land Imager (OLI) and the Thermal Infrared Sensor (TIRS) from which earth surface temperature could be extracted [10]. In this study, we used Landsat8 Level L1B data to calculate Land Surface Temperature (LST) to monitor the dynamics of Urban Heat Island (UHI).

2.2 Methodology

Exploratory data analysis was done is spreadsheets (Microsoft Excel) and the spatial analysis was done in s GIS (ArcGIS Desktop and Professional) environment. The analytical tools used were embedded in the programs and are elaborated below.

2.2.1 Spatial Autocorrelation

Moran's I statistic measures the spatial autocorrelation and is calculated as follows:

$$I = \frac{n \sum_{i,j} W_{ij} (Y_i - \overline{Y})(Y_j - \overline{Y})}{\sum_{i \neq j} W_{ij} \sum_i (Y_i - \overline{Y})^2} \tag{1}$$

where i and j were the region indexes and W_{ij} indicated the adjacency between area i and area j. The current study considered different types of adjacency in newly confirmed cases in areas i and j, respectively, and Y was the average of the number of newly confirmed cases in the entire region. A value of 0 indicated that there was no spatial autocorrelation in the data. A positive Moran's I value indicated the clustering of similar values, whereas a negative Moran's I value indicated the clustering of dissimilar values. The larger the absolute Moran's I value, the stronger the spatial autocorrelation [8].

2.2.2 Cluster Analysis

Hotspot Analysis

The Hot Spot Analysis tool calculates the Getis-Ord Gi* statistics for each feature in a dataset. The resultant z-scores and p-values tell you where features with either high or low values cluster spatially. This tool works by looking at each feature within the context of neighbouring features. High values are interesting but may not be necessarily be a statistically significant hotspot [1]. The hotspots of COVID-19 positive cases are mapped using Getis-Ord Gi* statistic algorithm to reveal the pattern of clusters where the disease breaks and spreads.

Cluster and Outlier Analysis (Anselin Local Moran's I)

Based on Moran's I statistic, the Cluster and Outlier Analysis tool identifies spatial clusters of features with high or low values. A positive value for I indicates that a feature has neighbouring features with similarly high or low attribute values; this feature is part of a cluster. A negative value for I indicates that a feature has neighbouring features with dissimilar values; this feature is an outlier. In either instance, the p-value for the feature must be small enough for the cluster or outlier to be considered statistically significant [2].

Spatial Interpolation

Ordinary kriging is a geostatistical interpolation method base on spatially dependent variance, which used to find the best linear unbiased estimate [4]. The general form of Ordinary kriging equation can be written as:

$$Z(x_p) = \sum\nolimits_{i=1}^{n} \lambda_i Z(x_i) \tag{2}$$

In order to achieve unbiased estimations in kriging the following set of equations should be solved simultaneously:

$$\sum\nolimits_{i=1}^{n} \lambda_i r(x_i, y_i) - \mu = r(x_i, x_p) where j = 1, \ldots, n \ with \sum\nolimits_{i=1}^{n} \lambda_i = 1 \tag{3}$$

where $Z(x_p)$ is the estimated value of variable Z at location x_p; $Z(x_i)$ is known value at location x_i; λ_i is the weight associated with the data; μ is the Lagrange coefficient; $r(x_i, x_j)$ is the value of variogram corresponding to a vector with origin in x_i and extremity in x_j; and n is the number of sampling points used in estimation.

LST Retrieval

Urban Heat Island is one of the most typical anthropogenic environmental characteristics in the urban area, During COVID-19 spread. The lockdown and 'stay-at-home' protocols changed the tracks and forms of human activity. Therefore, the urban environment is altered from both social and geographical aspects. To discover the

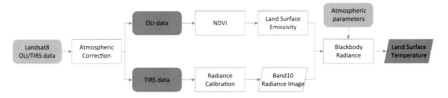

Fig. 2 Flowchart for LST retrieval

vague change of urban ecology under anti-pandemic regulations, the Land surface Temperature (LST) change between 2019 and 2020 was conducted.

Three surface temperature inversion algorithms: atmospheric correction method (also known as Radiative Transfer Equation, RTE), single-channel algorithm and split-window algorithm are currently reported in literature. For this study, the LST retrieval process was based on atmospheric correction method from the Landsat8 TIRS band (Fig. 2). The atmospheric effect on the surface thermal radiation was estimated first and then it was subtracted from the total amount of thermal radiation observed by the satellite sensor to obtain the surface thermal radiation intensity. It was then converted to the corresponding surface temperature.

The thermal infrared radiation brightness value L_λ received by the satellite sensor is composed of three parts: atmospheric upwelling radiation brightness $L^\uparrow$, the true radiation brightness of the ground reaches the energy of the satellite sensor after passing through the atmosphere and the energy reflected by the atmosphere as it radiates down to the ground [19].

By building Radiative Transfer Equation, the LST is obtained from the following expression:

$$L_{sensor,i} = \left[\varepsilon_i B_i(T_s) + (1 - \varepsilon_i)L^\downarrow_{atm,i}\right]\tau_i + L^\uparrow_{atm,i} \tag{4}$$

$$B_i(T_s) = \frac{L_{sensor,i} - L^\uparrow_{atm,i}}{\tau_i \varepsilon_i} - \frac{1 - \varepsilon_i}{\varepsilon_i}L^\downarrow_{atm,i} \tag{5}$$

$$T_s = \frac{c_2}{\lambda}[\ln\left(\frac{c_1}{\lambda^5 B_i(T_s)} + 1\right)]^{-1} \tag{6}$$

where L_{sensor} is the at-sensor radiance or Top of Atmospheric (TOA) radiance, i.e., the radiance measured by the sensor, $B_i(T_s)$ is the blackbody radiance of channel i, τ_i and ε_i are the atmospheric transmittance and land surface emissivity of channel i, and the $L^\uparrow_{atm,i}$ and $L^\downarrow_{atm,i}$ are the atmospheric upwelling and downwelling radiance of channel i, respectively, c_1 and c_2 are the Planck's radiation constants, for TIRS Band 10 data, the values is $774.89 \text{W}/\mu\text{m} \cdot m^2 \cdot sr$ and $1321.08\ K$, respectively, λ is wavelength. The atmospheric parameters τ_i, $L^\uparrow_{atm,i}$ and $L^\downarrow_{atm,i}$ can be calculated from in situ radio soundings and using a radiative transfer codes like MODTRAN.

3   Results

A statistical analysis and detection of environmental dynamics using measurable variables was done evaluate the correlation of COVID-19 and environmental consequences. The indicators chosen were typical factors that are affected by the consequences of pandemic spread. The qualitative analysis experiment was performed on the datasets to study the changing trend and statistical characteristics. The analysis results include the following visualised products:

- Maps of monthly distribution of COVID-19 cases.
- Graphs of daily concentration change of air pollutants.
- The temporal land surface temperature maps.

3.1   Spatial Analysis of Covid-19

3.1.1   Spatial Distribution in China and the US

Distribution Patterns

According to the to-date data of total cases in China and the United States, the spread of the epidemic spread in these two countries trends differently. China has controlled the epidemic spread within Hubei province since April 2020, whereas in the U.S. at the same time it was still spreading. Most cases in China concentrate in Hubei province and Hongkong (Fig. 3). Since Wuhan is the original outbreak city

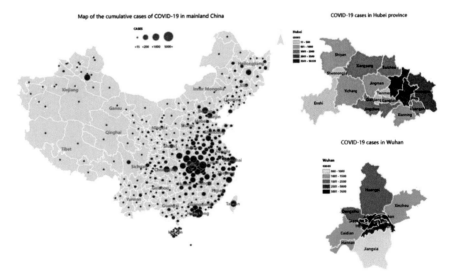

Fig. 3 COVID-19 total cases in mainland China, Hubei province and Wuhan City (Oct 10, 2020)

of COVID-19 in China, the death and positive cases mainly distributed in the urban area of Wuhan City.

In the case of the New York State in the U.S., the cases accumulated in several counties. New York City, as one of the foremost regions in the U.S. to be impacted by COVID-19, had large infections at the time (Fig. 4). In the comparative pie charts for China and the U.S. (Fig. 5), Wuhan and New York clearly stand out as the cities that's most impacted with Covid-19, at the time of this study.

Autocorrelation

The spatial autocorrelation of COVID-19 cases between the cities was estimated through Moran'I index by using the formula (1). The values of Moran coefficients are around the interval of [0, 1], since there is a positive correlation among the confirmed cases according to the geographical structure, and its spatial distribution has prominent agglomeration characteristics (Fig. 6). The increase of confirmed pneumonia cases in one region will inevitably lead to the increasing cases in adjacent

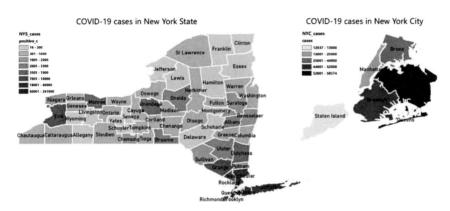

Fig. 4 COVID-19 total cases in New York State, New York City, U.S. (Oct 10, 2020)

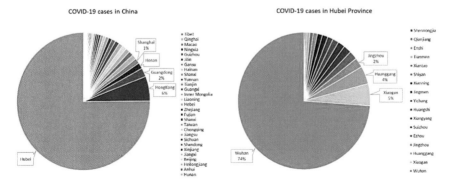

Fig. 5 COVID-19 Infection distribution

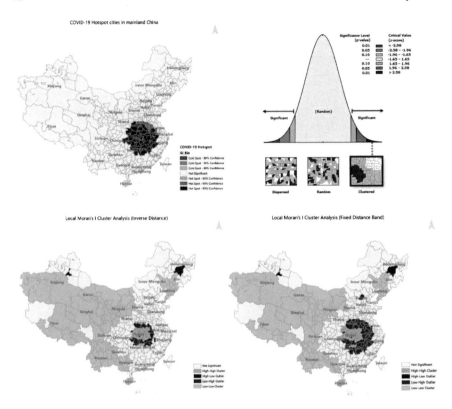

Fig. 6 Cluster analysis of COVID-19 cases in mainland China

areas, which means that a positive spill-over effect occurs. Thus, regions adjacent to COVID-19 hotspots are at higher *risk*.

3.1.2 Case Analysis in Wuhan City

The COVID-19 pandemic originated with a cluster of mysterious, suspected pneumonia cases in the city of Wuhan, the capital of Hubei, China in December 2019. The potential disease outbreak soon drew nationwide attention. On 23rd January 2020, the government announced lockdown order in Wuhan City. By 29th January, the virus spread to all provinces of mainland China [21]. The number of cases increased rapidly in February until the pandemic was under control in early March. By the time Wuhan ended lockdown order, over 2,575 died from the coronavirus infection-associated pneumonia, and 50,008 were confirmed to have been infected in Wuhan City (Fig. 7).

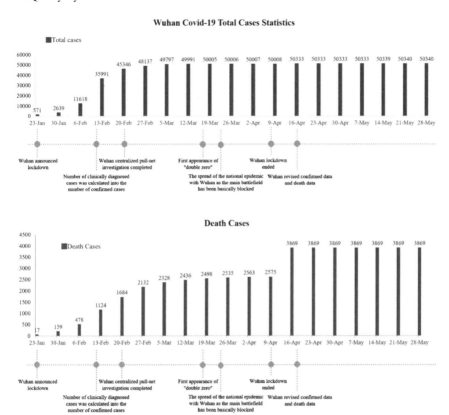

Fig. 7 Time series plot of newly confirmed COVID-19 cases, and death cases in Wuhan

The outbreak started from the urban districts and gradually spread to the suburban and rural areas across the disease pandemic periods. There were significant geographic differences in rates of confirmed cases, with the highest rates in the urban districts such as Jiangan, Jianghan, Qiaokou and Wuchang districts (Fig. 8).

3.1.3 Case Analysis in New York City

The first case relating to the COVID-19 pandemic was confirmed in New York City in March 2020. At the end of March, the infected cases increased to 33,983 at an unpredictable speed. By April, the city had more confirmed coronavirus cases than China, the U.K., or Iran, and by May, had more cases than any country other than the United States [13]. On 20 March, the governor's office issued an executive order closing non-essential businesses. The pause order caused the increasing unemployment rate and many social issues, whereas the infected cases of COVID-19 multiplied every day. The time series analysis of Covid-19 infections and deaths are presented

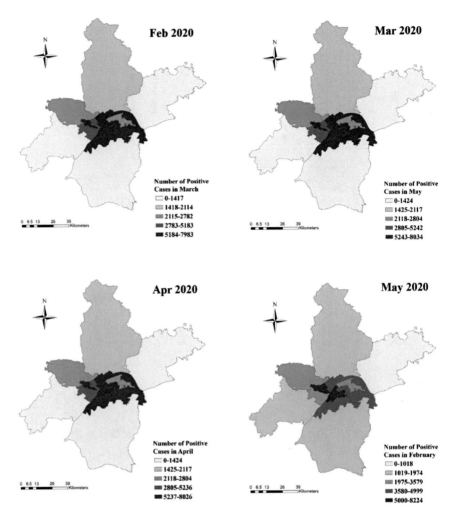

Fig. 8 Positive cases distribution in Wuhan (February, March, April, May)

in Fig. 9. When the city began its first phase of reopening on 8 June, the total cases had reached 211,728, accounting for 2% of the total population in NYC.

When infections counts were plotted per zip code, it was clear that some region had high number infection than others. To gain a better understanding, a hotspot analysis was conducted (Fig. 10) which showed Brooklyn, Bronx and the north of Queen Borough to be the most affected.

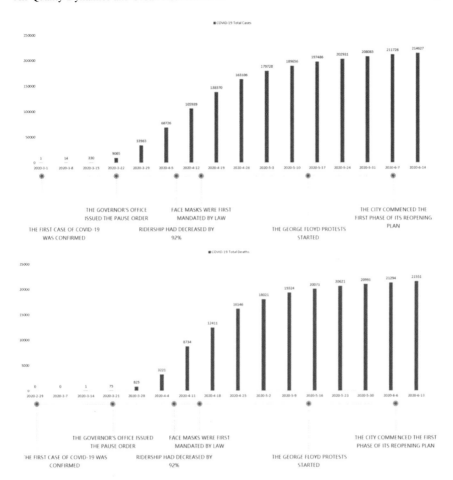

Fig. 9 Time series plot of newly confirmed COVID-19 cases, and death cases in New York City

3.2 Air Quality Dynamics

Vis-à-vis the temporal and spatial analysis of COVID-19 cases in two cities, environmental variables like air pollutants and land surface temperature were studied. Air pollution dynamics provides better understanding of retention of pollutants on the atmosphere which may contribute to spread of the virus. A month-wise analysis, for a span of six months, was done to compare the air quality during the COVID-19. Given that the time of the outbreak of the pandemic in the two cities was different, air quality data of Wuhan City and New York City was evaluated for a period between 1st January to 30th June. The environmental variables monitored were PM2.5, PM10, SO_2, NO_2, CO and O_3. The average air quality index (AQI), which is a dimensionless index, was calculated based on these six pollutants.

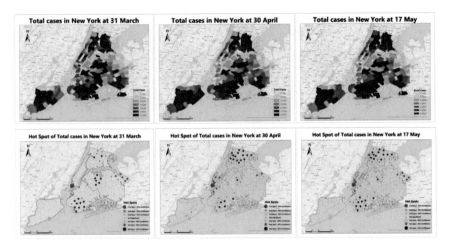

Fig. 10 Positive cases distribution based on zip-code area in NYC for March, April and May (Top) and corresponding Hotspots (Bottom)

3.2.1 Air Pollutants Distribution Patterns in Wuhan City

Figure 11 presents the daily concentration of the six criteria pollutants in Wuhan city from January 1st to April 30th. The daily average concentrations of $PM_{2.5}$ ranged from 8.5 $\mu g/m^3$ to 108.88 $\mu g/m^3$. In all 121 days, $PM_{2.5}$ in Wuhan City met the CAAQS Grade I standard (15 $\mu g/m^3$) for just 8 days and was in Grade II standard (35 $\mu g/m^3$) for 55 days. The better CAAQS standard (both I and II) was achieved in the period where the whole city was under quarantine. The daily average concentrations of PM_{10} ranged from 134.54 $\mu g/m^3$ to 12.88 $\mu g/m^3$. In 4 months, PM_{10} exceeded the Grade I standard (40 $\mu g/m^3$) for about 3 months and exceeded the Grade II standard (70 $\mu g/m^3$) for around 2 months. Monthly mean concentration of $PM_{2.5}$ decreased from 59.56 $\mu g/m^3$ to 34.11 $\mu g/m^3$ while monthly mean concentration of PM_{10} showed a U-shaped trend with the lowest point (69.82 $\mu g/m^3$) in February and the highest point (46.03 $\mu g/m^3$) in January. The high concentrations of $PM_{2.5}$ and PM_{10} reflected the fact that particulate matters were still the significant air pollutants during the COVID-19 epidemic.

The daily hourly concentrations of NO_2 exceeded the Grade I and II standard (40 $\mu g/m^3$, ~ 20 ppb) for 60 days, of which only 20 days in lockdown period (from January 23rd to April 8th). NO_2 showed an abrupt decline from January 23rd to February 1st and maintained low level from February to March. This was likely because of a series of stringent prevention and control measures issued by China government which resulted in a reduced number of vehicles plying on the road and diminished factory production [3, 24]. Due to the government's effectiveness of eliminating outdated production capacity and promoting clean energy, SO_2 is no longer the major problem [22]. SO_2 meet the Grade I standard (20 $\mu g/m^3$, ~7 ppb) for 103 of 121 days, and meet the Grade II standard (60 $\mu g/m^3$, ~ 21 ppb) for

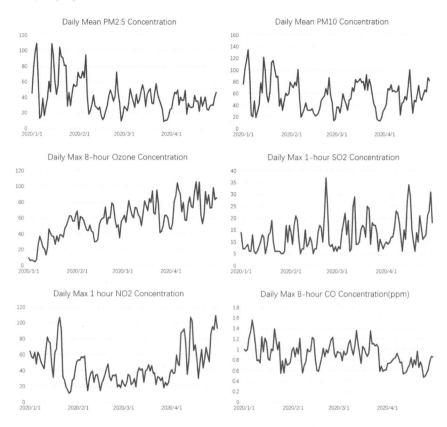

Fig. 11 Daily mean concentrations of air pollutants PM2.5 (μg · m^{-3}), PM10 (μg · m^{-3}), O$_3$(μg · m^{-3}), SO$_2$ (μg · m^{-3}), NO$_2$ (μg · m^{-3}), and CO (ppm) from 1st January to 31st June in Wuhan City

all 4 months. Daily average 8 h O$_3$ concentrations ranged from 105.46 μg/m^3 to 4.46 μg/m^3, and daily average CO concentrations ranged from 1.56 ppm to 0.47 ppm. The large quarantine in China contributed to a significant decline in concentration of air pollutants except O$_3$. Unlike the other 5 pollutants, the concentration of O$_3$ showed an upward trend, increased from 33.34 μg/m^3(January) to 78.78 μg/m^3 (April). This was most likely because of high intensity of solar radiation which converts oxygen into ozone evidenced by the reverse seasonal variation observed in 8 hourly O$_3$ concentrations in the summer and the lowest in the winter [18].

To visually demonstrate of the air pollution variation, a map tool in ArcGIS Pro to graphically depict the distribution of pollutants concentrations in each of the seventeen Wuhan districts, was used. The spatial–temporal distribution of different air pollutants was significantly heterogeneous among these districts (Figs. 12 and 13). Specifically, the concentrations of five air pollutants during the lockdowns appeared to be much lower than concentrations on regular days, which offered supportive evidence of the pollution reduction effects of travel restrictions. The concentrations

CO distribution in WuHan City from January to June 2020

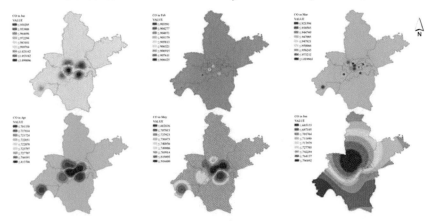

NO2 distribution in WuHan City from January to June 2020

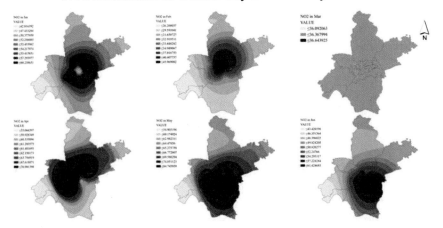

Fig. 12 Distribution of monthly mean concentrations NO_2 ($\mu g \cdot m^{-3}$), and CO (ppm) and PM2.5 ($\mu g \cdot m^{-3}$) from 1st January to 31st June in Wuhan City

of CO and NO_2 have the geographical feature of "low outside and high inside". JH and WC, which belong to the central urban area of Wuhan City, were the area with the most serious CO and NO_2 pollution because of developed traffic and large traffic volume. PM2.5 and PM10 are the highest in QS District because the petroleum and power generation industries in QS District are very developed. Unlike CO and NO_2, the concentration of O_3 and SO_2 at rural area is higher than that at central area. Especially, high concentrations for O_3 and SO_2 were noticed in the northern part and southern part respectively. This is because the high vegetation rate of suburbs can release a large number of VOCs, which is conducive to O_3 production. At the

PM2.5 distribution in WuHan City from January to June 2020

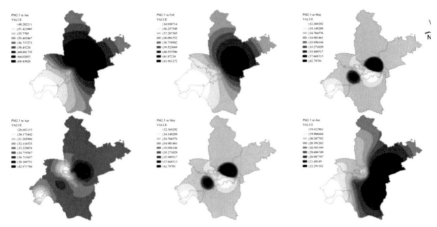

Fig. 12 (continued)

same time, the NO emitted by vehicles can react with O_3, which is equivalent to consuming O_3 to a certain extent.

3.2.2 Air Pollutants Distribution Patterns in New York City

In NYC air quality analysis, the PM10 data was missing in most sites. Therefore, the evaluation covered only five criteria air pollutants. According to Fig. 13, the daily variations of five air pollutants present different change trend. The concentration of fine particulate matter (PM2.5) shows a tendency to decrease in the first three months and then increase. The Ozone concentration slightly goes up from 1st March while the nitrogen dioxide (NO_2) gradually decreases during the pandemic outbreak. The changes of SO_2 and CO are not significant, except some peak values occasionally occurs in May.

The monthly distribution maps of air pollutants are based on the mean values of concentration from all in-situ monitoring sites within and around the NYC urban area. The box charts (Fig. 14) show the variation of the monthly average of PM2.5, Ozone, SO_2, NO_2, CO. The lowest point (2.28 $\mu g \cdot m^{-3}$) of PM2.5 appears in May, two months after the 'Pause Order'. The Ozone concentration ranges from 0.027 $\mu g \cdot m^{-3}$ to 0.044 $\mu g \cdot m^{-3}$, following the timeline. SO_2, NO_2 and CO present a decreasing trend from January 2020 to June 2020. The lowest point of SO_2 is 0.64 $\mu g \cdot m^{-3}$ in April while the highest point is 1.12 $\mu g \cdot m^{-3}$ in January. The NO_2 concentration varies from 22.55 $\mu g \cdot m^{-3}$ in May to 33.98 $\mu g \cdot m^{-3}$ in February, with the similar temporal variation patterns to CO. The value of CO concentration reaches the highest point (0.505 ppm) in February and lowest point in May (0.247 ppm).

PM10 distribution in WuHan City from January to June 2020

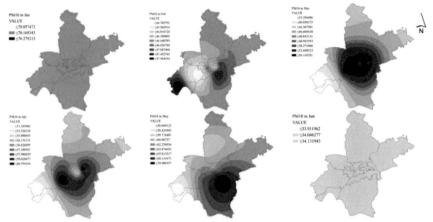

SO2 distribution in WuHan City from January to June 2020

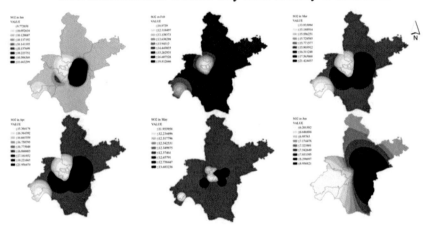

Fig. 13 Distribution of monthly mean concentrations of PM10 ($\mu g \cdot m^{-3}$), SO$_2$ ($\mu g \cdot m^{-3}$) and O$_3$ ($\mu g \cdot m^{-3}$) from 1st January to 31st June in Wuhan City

The spatial distribution of five criteria pollutants in New York City from January 2020 to June 2020 is provided from Figs. 15, 16, 17, 18 and 19. The spatial features of pollution distribution are closely related to urban configuration and land use type. The commercial zones mostly located in Manhattan borough and industrial & manufacturing facilities are separated along the shoreside in Staten Island. The northern and eastern part of New York City like Bronx, Queens and Brooklyn are mostly residential areas.

The eastern part of NYC had lower PM2.5 concentration from February to May in 2020. After the first phase reopening order released, the PM2.5 in residential areas went higher than surrounding areas. Daily temperatures, relative humidity, and wind

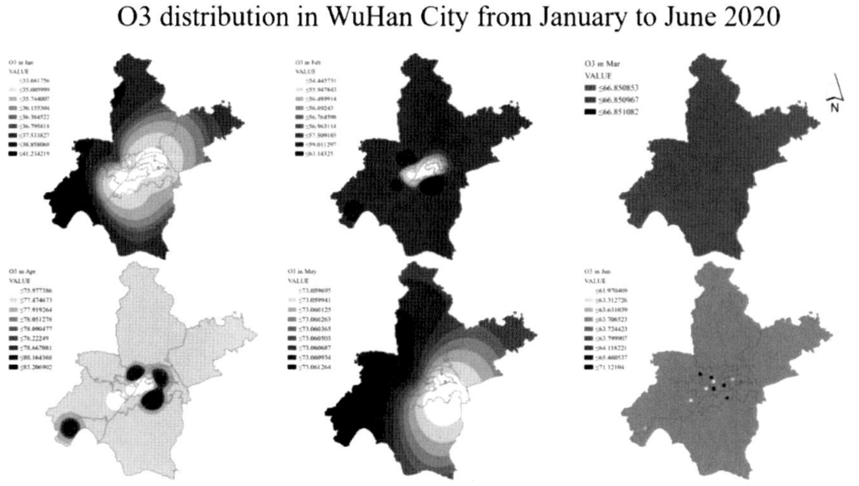

O3 distribution in WuHan City from January to June 2020

Fig. 13 (continued)

speed can affect ozone levels. In general, warm, dry weather is more conducive to ozone formation than cool, wet weather. NOx and VOC emissions can also influence ozone levels. The distribution of ozone displayed similar spatial patterns in March and April. In June, the center of high-density ozone moved to eastern part in the urban area.

SO_2, NO_2 and CO primarily get in the air from the burning of fuel. NO_2 forms from emissions from cars, trucks and buses, power plants, and industrial facilities. In NYC, the high concentration of SO_2 in the atmosphere clustered in the queen borough from January to May 2020. After the pandemic outbreak, the NO_2 and CO pollution concentrated in Staten Island borough where most industrial facilities located.

3.3 Urban Heat Island (UHI) Effects

To evaluate the impacts of COVID-19 on the urban thermal environment, the temporal maps of LST distribution are produced to make a comparison between 2019 and 2020. Generally, the UHI effect is more significant in summer because of the higher temperature and more CO2 emissions generated from air-conditioning usage.

Wuhan City, known as one of the 'Stove City' in China, has obvious UHI effects in central urban area. From Fig. 20, the distribution patterns of LST are similar in February and March. The rural area has a higher temperature than the urban area, which is assumed as a cold island effect caused by the difference in heat emissions in the urban and rural area. Through the comparison of LST in August 2019 and 2020, we found that the UHI effect is severe in 2020 (Fig. 21). Apart from the influence of

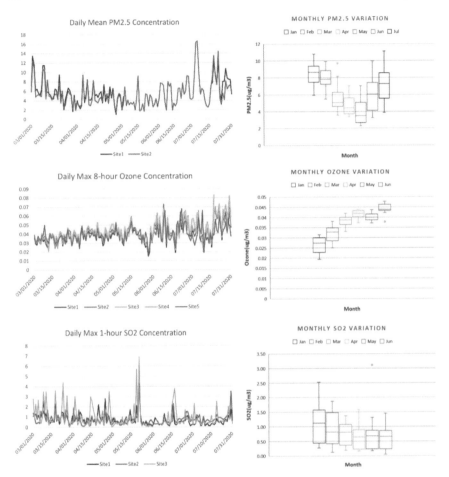

Fig. 14 Concentration variation of five air pollutants: PM2.5 ($\mu g \cdot m^{-3}$), O_3($\mu g \cdot m^{-3}$), SO_2 ($\mu g \cdot m^{-3}$), NO_2 ($\mu g \cdot m^{-3}$), and CO (ppm) in New York City

global warming, the massive work resumption after months' lockdown is likely to be one of the main causes for the increase of temperature in the urban area in Wuhan City.

Contrary to Wuhan City, the UHI effect in New York City seems to mitigate in 2020 compared with the same time in 2019. Previous studies show that the epidemic influenced normal human activities. The decreased flow in the industrial and commercial area caused lower emission of CO_2. Although the change is not significant, it is safe to conclude that the COVID-19 affected the environmental variations in Wuhan City and New York City.

The Moran's I index of LST point features, exhibits a clear clustering of LST (Table 3). The temperature in each location affects the adjacent location. More

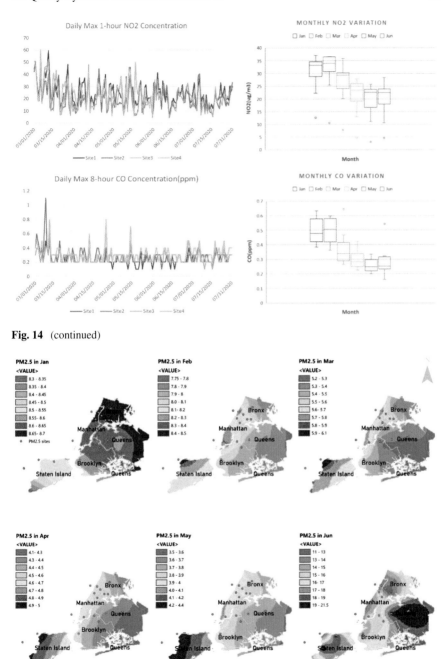

Fig. 14 (continued)

Fig. 15 Spatial interpolation map of monthly mean concentrations of PM2.5 (μg · m^{-3}) from 1st January to 31st June in New York City

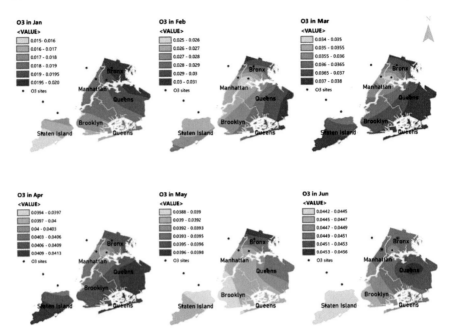

Fig. 16 Spatial interpolation map of monthly mean concentrations of O_3 ($\mu g \cdot m^{-3}$) from 1st January to 31st June in New York City

analysis can be conducted to study the range of the inner influence between the geographical areas in the megacities.

4 Discussion

The analysed spatial data in this study demonstrates the impact of COVID-19 on environmental variations in the two epi-centre cities. The illustrations of dynamic changes of air pollution level and urban heat island effects in Wuhan and New York City provide information for the study of socio-ecological consequences of an epidemic outbreak. In conclusion, we found that the restrictions on human activities have positive interactions with the urban environment. The spatial–temporal patterns of air pollutants distribution were modified, especially in Wuhan City. The air quality reached the standard level due to the lockdown order. However, the resumption of industrial production and commercial activities after June 2020 brought a more severe UHI effect in central urban areas. For New York City, air pollution decreases during the outbreak time, but the change is not as significant as in Wuhan City. In addition, the UHI effects in NYC alleviated in 2020 compared with the same time in 2019. The factors contributing to the weaken UHI effects are complicated. Given the global

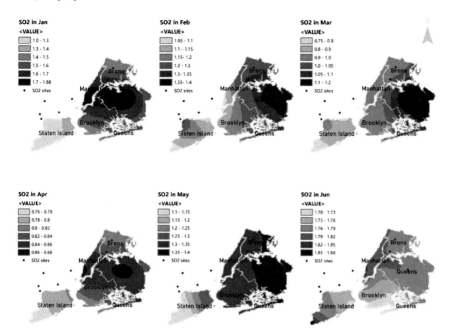

Fig. 17 Spatial interpolation map of monthly mean concentrations of SO_2 ($\mu g \cdot m^{-3}$) from 1st January to 31st June in New York City

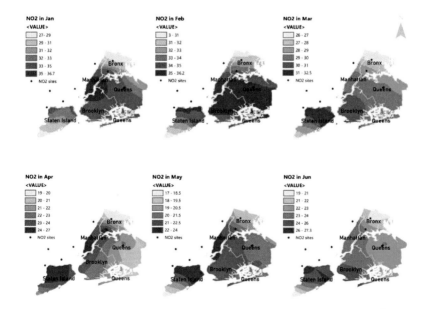

Fig. 18 Spatial interpolation map of monthly mean concentrations of NO_2 ($\mu g \cdot m^{-3}$) from 1st January to 31st June in New York City

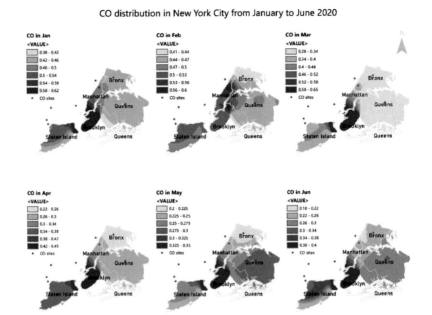

Fig. 19 Spatial interpolation map of monthly mean concentrations of CO (ppm) from 1st January to 31st June in New York City

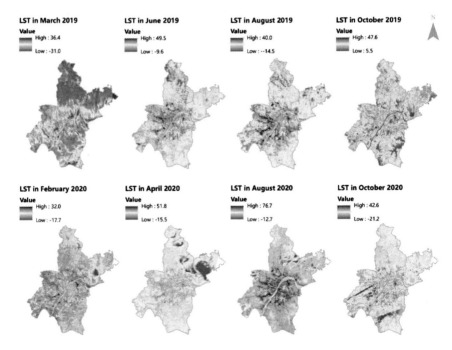

Fig. 20 Land surface temperature maps of Wuhan

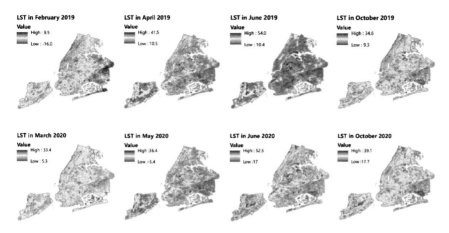

Fig. 21 Land surface temperature maps of New York City

Table 3 Moran's I index of the land surface temperature

City	Month	Moran'I	Z-score	P-value
New York City	March	0.2433	7.733	< 0.01
	May	0.3114	10.0953	< 0.01
	June	0.1878	5.9727	< 0.01
	October	0.2972	9.3759	< 0.01
Wuhan City	February	0.4239	13.0152	< 0.01
	April	0.6438	20.0957	< 0.01
	August	0.4554	14.4435	< 0.01
	October	0.6606	20.9927	< 0.01

warming tendency, the responses of the local community to COVID-19 are likely to be one of the social factors that influenced the urban thermal environment.

The limitation and future improvements envisaged on the present study are as follows:

1. The correlation analysis between COVID-19 cases and air pollutants concentration was not conducted due to the limited data. With the available long-term data, one can make a more valid analysis of air quality dynamics in the pandemic period.
2. Socio-economic data and human mobility data are supposed to be prompt critical elements for the investigation of epidemic influence. In future study, such data could be taken into consideration.
3. The air and thermal factors are inseparable parts for the environmental assessment of urban context, which directly affects public health and economic development. The relationship between these two kinds of factors is another meaningful topic to study.

The current study provides insights for the research of environmental consequences of COVID-19 on a local scale. Despite the initial finding of this analysis,
more correlations between critical environmental variables and disease spread can
be explored in the future. Since the adaptation to long-lasting coronavirus crisis will
be a new normal for human society, the statistical and quantitative analysis should
be applied to assist the policymaking, economic renewal, and urban environmental
restoration.

References

1. Andy M (2005) The ESRI guide to GIS analysis, volume 2: spatial measurements and statistics
2. Anselin L (1995) Local indicators of spatial organization -LISA. Geogr Anal 27(2):93–115
3. Bao R, Zhang A (2020) Does lockdown reduce air pollution? Evidence from 44 cities in northern
   China. Sci Total Environ 731:139052. https://doi.org/10.1016/j.scitotenv.2020.139052
4. Belkhiri L, Tiri A, Mouni L (2020) Spatial distribution of the groundwater quality using kriging
   and Co-kriging interpolations. Groundw Sustain Dev 11:100473. https://doi.org/10.1016/j.gsd.
   2020.100473
5. Cordes J, Castro MC (2020) Spatial analysis of COVID-19 clusters and contextual factors in
   New York City. Spat Spat-Temporal Epidemiol 34:100355. https://doi.org/10.1016/j.sste.2020.
   100355
6. Franch-Pardo I, Napoletano BM, Rosete-Verges F, Billa L (2020) Spatial analysis and GIS in
   the study of COVID-19. A review. Sci Total Environ 739:140033. https://doi.org/10.1016/j.sci
   totenv.2020.140033
7. Huang R, Liu M, Ding Y (2020) Spatial-temporal distribution of COVID-19 in China and its
   prediction: a data-driven modeling analysis. J Infect Dev Ctries 14(3):246–253. https://doi.org/
   10.3855/jidc.12585
8. Kang D, Choi H, Kim JH, Choi J (2020) Spatial epidemic dynamics of the COVID-19 outbreak
   in China. Int J Infect Dis 94(January):96–102. https://doi.org/10.1016/j.ijid.2020.03.076
9. Li H, Calder C, Cressie N (2007) Beyond moran's i: testing for spatial dependence based
   on the spatial autoregressive model. Geogr Anal 39:357–375. https://doi.org/10.1111/j.1538-
   4632.2007.00708.x
10. Li RYM, Chau K, Li H, Zeng F, Tang B, Ding M (2021) Remote sensing, heat island effect and
    housing price prediction via autoML (pp 113–118). https://doi.org/10.1007/978-3-030-51328-
    3_17
11. Mukherjee S, Debnath A (2020) Correlation between land surface temperature and urban heat
    Island with COVID-19 in New Delhi, India, pp 1–11. https://doi.org/10.21203/rs.3.rs-30416/v1
12. Nakada LYK, Urban RC (2020) COVID-19 pandemic: impacts on the air quality during the
    partial lockdown in São Paulo state, Brazil. Sci Total Environ 139087
13. NPR (2020) March 24, 2020. New York City, U.S. Epicenter, Braces For Peak
14. Peng Z, Wang R, Liu L, Wu H (2020) Exploring urban spatial features of COVID-19 transmis-
    sion in Wuhan based on social media data. ISPRS Int J Geo-Inf 9(6). https://doi.org/10.3390/
    ijgi9060402
15. Rodríguez-Urrego D, Rodríguez-Urrego L (2020) Air quality during the COVID-19: PM2.5
    analysis in the 50 most polluted capital cities in the world. Environ Pollut 266:115042. https://
    doi.org/10.1016/j.envpol.2020.115042
16. Ruiz Estrada M (2020) economic waves: the effect of the wuhan COVID-19 on the world
    economy (2019–2020). https://doi.org/10.13140/RG.2.2.11861.99047/1
17. U.S. Census Bureau QuickFacts: New York City, New York. (2020) https://SO<Subscript>2<
    Subscript>ewyork. Accessed 28 May, 2020

18. Wang Y, Ying Q, Hu J, Zhang H (2014) Spatial and temporal variations of six criteria air pollutants in 31 provincial capital cities in China during 2013–2014. Environ Int 73:413–422. https://doi.org/10.1016/j.envint.2014.08.016
19. Wang M, Zhang Z, He G, Wang G, Long T, Peng Y (2016) An enhanced single-channel algorithm for retrieving land surface temperature from Landsat series data. J Geophys Res: Atmos 121(19):11712–11722. https://doi.org/10.1002/2016JD025270
20. Yang Q, Huang X, Tang Q (2019) The footprint of urban heat island effect in 302 Chinese cities: temporal trends and associated factors. Sci Total Environ 655:652–662. https://doi.org/10.1016/j.scitotenv.2018.11.171
21. Yuan L (2020) Coronavirus crisis exposes cracks in China's facade of unity. The New York Times. ISSN 0362-4331
22. Zhang H, Di B, Liu D, Li J, Zhan Y (2019) Spatiotemporal distributions of ambient SO2 across China based on satellite retrievals and ground observations: substantial decrease in human exposure during 2013–2016. Environ Res 179:108795. https://doi.org/10.1016/j.envres.2019.108795
23. Zhang X, Estoque RC, Murayama Y (2017) An urban heat island study in Nanchang City, China based on land surface temperature and social-ecological variables. Sustain Cities Soc 32(May):557–568. https://doi.org/10.1016/j.scs.2017.05.005
24. Zhang X, Tang M, Guo F, Wei F, Yu Z, Gao K, Jin M, Wang J, Chen K (2021) Associations between air pollution and COVID-19 epidemic during quarantine period in China. Environ Pollut 268:115897. https://doi.org/10.1016/j.envpol.2020.115897

Mass Rapid Transit and Population Dynamics During Covid-19 in Singapore

Gu Qianhua, Li Ruoyu, Wang Jing, and Zou Hongyi

Abstract Mass Rapid Transit (MRT) is the backbone of Singapore's transport system. It ferries its population from their homes to work and back. On weekends, people use the MRT to visit places of leisure. This travel ritual of Singapore residents was disrupted during the onset of Covid-19 where strict measures were enforced to contain the infection chain. These measures, called circuit breakers, were enforced and later relaxed to allow people to carry out some activities which involved MRT travel. From 2nd of June 2020 the country entered 'Phase One' of safe re-opening during which certain economic activities, which do not pose high risk of transmission, were allowed and the remaining ones, with higher risk of disease transmission remain closed. Ridership of MRTs partly recovered during this period. 'Phase Two' of re-opening followed from 19th June which allowed selective recreational activities with safe distancing measures still in place. The current study was aimed to understand the Spatio-temporal patterns of MRT ridership during Phase One and Two. The study reflects on the MRT stations which were popular and brings forth insights which may help city planning authorities for the future.

1 Introduction

Singapore went into a lockdown to break the chain of Covid-19 in a phased manner. The lockdown was not totalitarian, and people could travel albeit safe distancing rules. The restrictions impacted several aspects of public transport and new patterns emerged from movements of people. MRTs being the backbone of the country's transport system, was chosen as the subject of the study as high-density temporal data for ridership was being acquired and stored by the government and was available for the study.

G. Qianhua · W. Jing (✉) · Z. Hongyi
4 Architecture Dr, Singapore 117566, Singapore

L. Ruoyu
701, Gate 3, Building 4, Area 54A, Jilin province, Changchun 130000, China

© The Author(s), under exclusive license to Springer Nature Singapore Pte Ltd. 2022
S. N. Kundu (ed.), *Geospatial Data Analytics and Urban Applications*,
Advances in 21st Century Human Settlements,
https://doi.org/10.1007/978-981-16-7649-9_4

1.1 MRT Ridership and Urban Development

Mass rapid transit (MRT) is the backbone of public transport system in Singapore. The first MRT section was opened in November 1987 [8]. According to [4], the rail network is planned to expand to about 360 km by 2030 from the current 200 km. The expansion aims to connect 8 in 10 households to within 10 min of an MRT station.

Ridership, or passenger volume, is one of the most used measure to capture the effect of the surrounding land use, clustered development, diversity, density, transit supply and system efficiency on transit use [2]. Higher ridership also reveals higher popularity and better economic benefits.

Land use allocation influences the popularity of MRT stations. As Zhu et al. (2004) mentions, the existing urban land use configuration helps to shape travel patterns. In Singapore, the urban center hierarchy and the new town development concept have led to the difference in land use characteristics of the Transit-oriented Development (TOD) stations [7]. Many studies correlate ridership with land use. [6] reviewed related papers and concluded that high public transit ridership was related to high land use density. In addition, diversity of land uses among three major categories within the walking radius of station can also accumulate transit passenger volume [3] used transit smart card data to identify the travel pattern and reveal the relationship between travel patterns and the surrounding environments in Seoul. Through quantifying linear functions consisting of ridership and GFA, [1] found that office development can generate more passengers while residential development can generate less.

1.2 Influence of Covid-19 Restrictions

Pandemic also influences the ridership of public transport. Because of COVID-19 outbreak, the transport sector experienced a drastic reduction in passenger traffic, including MRT system [5]. According to LTA (2020), ridership of MRT plummeted by 75 per cent in April as compared to pre-COVID levels in Singapore.

Singapore ended the Circuit Breaker from 2 June 2020 and entered Phase One of safe re-opening. Economic activities, which did not pose high risk of disease transmission, were gradually re-opened in this period, while social, economic and entertainment activities with higher risks remain closed. Passenger flow of MRT partially recovered as compared to the Circuit Breaker period. Phase Two of re-opening started from 19 June 2020. Certain places for recreation and activities gradually reopened in the late June or early July under strict safe distancing measures leading to the increase in MRT ridership in July.

The spatial analysis of MRT ridership helps understand human mobility which in turn provide insights into daily human activities. Such studies can help improve the role of MRT stations in society as underlying urban factors are better understood. The identification of popular MRT stations, during restricted easing of travel during pandemics provides an opportunity to discourse reasons for its popularity which is turn paves the way more careful and effective urban planning.

2 Objectives, Data and Methods

2.1 Objectives

The current research had multiple objectives which can listed as below.

- Measure the passenger volume of each destination MRT station post Covid-19 restrictions.
- Discover variations in passenger volume trends for weekdays and weekends.
- Correlate urban functions with the discovered trends and variations.
- Identify and reason the unusual popularity of certain destination MRT stations.

2.2 Data and Sources

The data for this research were sourced from multiple portals primarily run by the Government of Singapore. These portals extend an API (application processing interface) to interface desktop based spatial data processing platforms such as Geographical Information Systems. The datasets used for the study were:

- Monthly passenger volume by origin–destination MRT train stations was sourced from Land Transport Data Mall. (https://www.mytransport.sg/content/mytransport/home/dataMall.html).
- The Singapore Master Plan 2019 Land Use GIS data from Department of Statistics Singapore (https://www.singstat.gov.sg) was used for land use types. This also included spatial locations for the MRT network and stations.

2.3 Framework

The framework for the research included several steps for preparing the data. The first step was exploratory analysis and preparation for the data for next level analytics (Fig. 1). This included Data wrangling, Data Manipulation, Spatial overlay analysis, Statistical summarizing, and typological classification.

The next phase was the analytical framework (Fig. 2), where popularity analysis and popularity variation analysis were conducted, and statistical outputs were visualised as tables and maps.

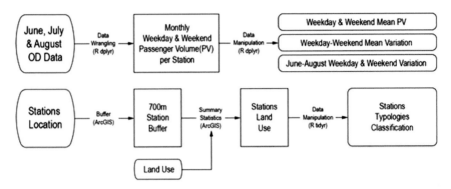

Fig. 1 Data Preparation Framework for the study

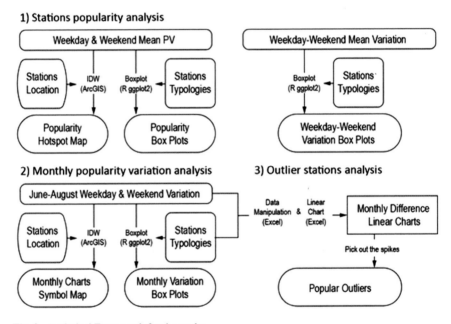

Fig. 2 Analytical Framework for the study

2.4 Algorithms Used

One essential ingredient for any data science project are the algorithms. These algorithms are not only critical but also crucial to the inferences drawn on the subject. The algorithms used were mostly statistical and are defined as below

$$WEEKDAY\ MEAN\ PV = \frac{average\ PV\ on\ weekdays}{total\ number\ of\ weekdays} \tag{1}$$

$$WEEKEND\ MEAN\ PV = \frac{average\ PV\ on\ weekends\ \&\ holiday}{total\ number\ of\ weekends\ \&\ holiday} \quad (2)$$

$$WEEKDAY - WEEKENDVARIATION =$$
$$\frac{WEEKENDMEANPV - WEEKDAYMEANPV}{WEEKDAYMEANPV} \quad (3)$$

$$JUNE - AUGUSTWEEKDAYVARIATION =$$
$$\frac{AUGUSTWEEKDAYMEANPV - JUNEWEEKDAYMEANPV}{JUNEWEEKDAYMEANPV}$$
$$\quad (4)$$

$$JUNE - AUGUSTWEEKENDVARIATION =$$
$$\frac{AUGUSTWEEKENDMEANPV - JUNEWEEKENDMEANPV}{JUNEWEEKENDMEANPV}$$
$$\quad (5)$$

3 Analysis and Results

3.1 Typology Classification of MRT Stations

The MRT station locations follow the urban planning principles. The surrounding urban land use around each MRT station specify the urban functions that the station serves and also defines the MRT user groups which start and end their travel at that particular station. This provided the basis for classification of MRT stations into several typologies (Fig. 3). A distance buffer of 700 m was used around each MRT station to extract and summarize the land use. In Singapore, a distance of 700 m is considered as a typical walkable distance.

Land use types were extracted for a circular buffer area of radius 700 m around each MRT station. 7 station typologies based on urban planning concepts were used for this study. These were.

Changi Airport   Harbourfront   Botanic Gardens   Bayfront   Marina South Pier

Fig. 3 Land use with 700 m zone of some MRT stations

1. **Residential**: The station is fully surrounded by residential housing, possibly with a small amount of educational land, and most passengers are neighborhood residents.
2. **Residential with Town Center:** The station also serves the nearby community, but there are commercial developments around the station that form a lively town/regional centre (e.g. Ang Mo Kio and Clementi MRT Stations).
3. **Open Space**: The station is adjacent to park, open space or waterbody, which usually acts as recreation destination (e.g. Botanic Gardens and Harbourfront MRT stations).
4. **Commercial**: The station is surrounded by a large commercial area and deemed to be an important shopping destination and office space in the city (e.g. Orchard and Bugis MRT stations).
5. **Business**: The station is surrounded by industrial & business parks, concentrating major manufacturing industries (e.g. Jurong East and One-North MRT stations).
6. **Public Facilities**: The station is located close to public facilities such as schools, stadiums, transportation ports etc. Most of the passengers alighting here are likely users of these facilities (e.g. Changi Airport, Stadium).
7. **Others**: The site locates in mixed-use urban area, or undeveloped reserve site, or military base.

After developing the typologies, several representative stations under each typology were selected and the land use proportion within their 700 m buffers extracted as benchmark to develop the criteria for classifying the stations (Figs. 4 and 5).

According to this classification criteria, all 158 stations were sorted under different typologies, resulting 53 Residential Stations, 21 Residential with Town Center Stations, 26 Open Space Stations, 15 Commercial Stations, 23 Business Stations, 13 Public Facilities Stations and 7 other Stations (Fig. 5).

Fig. 4 Station Typologies

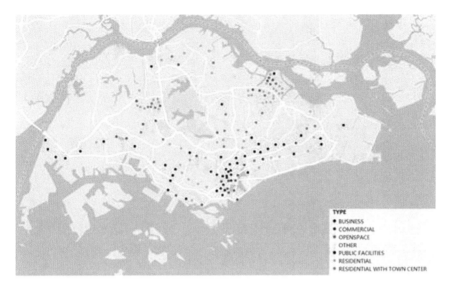

Fig. 5 Spread of MRT Stations in Singapore (colour indicates typology)

3.2 Popularity Analysis

Popularity analysis was based on several tools which include hotspot analysis, statistical boxplots and variation boxplots.

3.2.1 Hotspot Analysis

Hotspots for weekdays and weekends were generated for the study period. The two hotspot maps (Fig. 6) were studied for visual differences. Some stations appeared to be popular during the weekends e.g. Tanjong Pagar, Tampines, Rochor and Jalan Besar. Etc, as these were located at popular commercial destinations where people flock during weekends for purchases.

The weekdays map distinctly displays four hotspots in Singapore around the Jurong industrial estate, Yishun, CBD and Changi. These are where most commercial and business establishments are located where people needed to travel during weekdays. Six MRT station service these locations.

3.2.2 Boxplots

Boxplots are an effective tool for visualization of statistical information. They intuitively display the minimum, the maximum, the sample median, the first and third quartiles, interquartile range, and outliers in a single diagram (Fig. 7). The size,

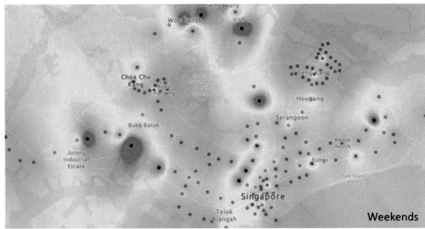

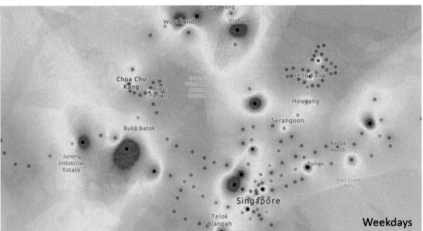

Fig. 6 Hotspot Analysis for weekends (top) and weekdays (bottom)

concentration, and abnormal value of Passenger Volume (PV) according to different types of MRT stations in weekdays and weekends, as well as the variation in popularity of different types of MRT stations from weekdays to weekends could be visualized using boxplots.

The median value was an indicator or MRT popularity for different types of land use. E.g. Residential with town center and commercial area were found to have high median values as they are popular on weekdays. A low median value of PV in a business land use was possibly affected by non-operational industrial areas (which were defined as business land use according to 2019 master plan) with quite low passenger volume.

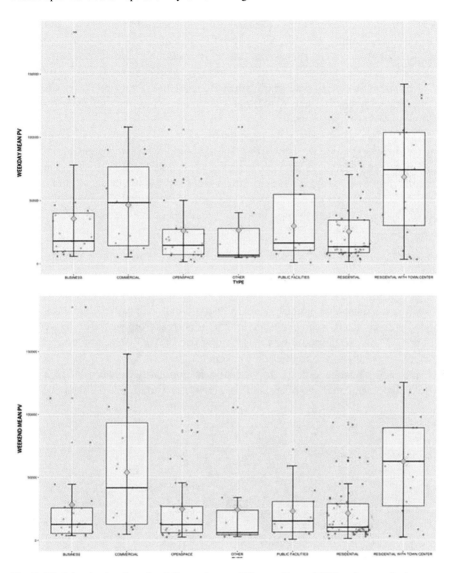

Fig. 7 Weekday (top) and weekend (bottom) mean PV according to MRT station typologies

Interquartile range reflects the variation in popularity of different MRT stations belonging to same typologies. Passenger volume in residential with town center varied the most, followed by commercial areas and public facilities.

Business land use gets two distinct outliers and one of them in the Jurong Industrial estate. This is where manufacturing businesses are hosted where employees are most likely unable to work from home and needed to travel to office to execute their job tasks. Open spaces also presented outliers which more distinct as design and development of different open spaces in Singapore are different to industrial hubs. For instance, well-developed open spaces like botanic garden are more popular to public while other green spaces that lack in recreational value did not attract more passenger volume to those locations. Variations in open spaces may also be related to their different reopening status during the post-pandemic. Residential hubs displayed the largest number of outliers with limited variability which was mostly linked to the different types of housing.

The median value indicated a significant increase in popularity of commercial areas from weekdays to weekends (Fig. 7). Residential with town center and commercial areas continue to be more popular during weekends. The variation in popularities of commercial areas shows an increase on weekends, while business land use and public facilities shows distinct decrease. Possible reasons are the decrease of working population and the increasing commercial and leisure activities like shopping on weekends.

3.2.3 Variation Plots

The high median value indicates a sharp increase in popularity of commercial areas and significant decrease in popularity of business areas from weekdays to weekends (Fig. 8). Main reason could be the declining use of MRT for commuting and the increasing commercial activities on weekends.

Interquartile range reflects the variation in popularity between weekdays and weekends. Residential and other land use show little variation, which may be due to

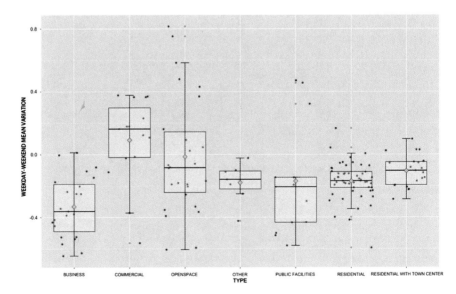

Fig. 8 Weekday-weekend mean variation boxplots

the stable daily activities and use of MRT near passengers' home. Open space varies most with two observable outliers. Main reason could be the increase of leisure activities based on reopening status of different open spaces. Different design and development levels also influence the variation and outliers of open space.

Besides open space, public facilities also present distinct outliers. Possible reason could be the increasing traveling during weekends, which lead passengers to places like Changi Airport.

3.3 Popularity Patterns and Variations

Monthly popularity patterns were generated and studies. Variations observed were analyses and reasoned.

3.3.1 Monthly Patterns

In general, weekdays' passenger flow volume of the most MRT stations are growth at a constant speed with the pie charts at almost three-equal (Fig. 9). In detail, the pie chart map shows significant differences at 3 MRT station. One pie chart shows there has no passenger flow volume on June, has a little bit passenger flow volume on July and considerably increased in August. The other two have a more stable increase than the first one, but still shows a rapidly rise compare to the others.

The variation pattern of passenger flow on weekends is somewhat different from that on weekdays, some MRT stations have the highest variation in July, but they decreased rapidly in August.

3.3.2 Monthly Variations

The median illustrates higher increase in the variation in popularity of commercial areas and open spaces during weekdays from June to August, while the variation in business areas is relatively low (Fig. 10). Main reason could be the reopening status of different land uses in Phase One and Phase Two. Business first recovered from the Circuit Break from 2nd June 2020, while part of commercial activities and open spaces gradually reopened after 19th June (most of which reopened in July).

In terms of the increase in popularity of MRT stations with different typologies from June to August, which is indicated by interquartile range, open spaces vary most, followed by public facilities. Both typologies show distinct outliers. The variation and outlier of open space are possibly caused by different reopening status and the development of each open space. The outlier of public facility (Changi Airport) is mainly caused by the reopening of transit and commercial activities in July.

Overall, the increase of popularity of each type of land use during weekends is more obvious than that of weekdays. Main reason could be the recovering of human

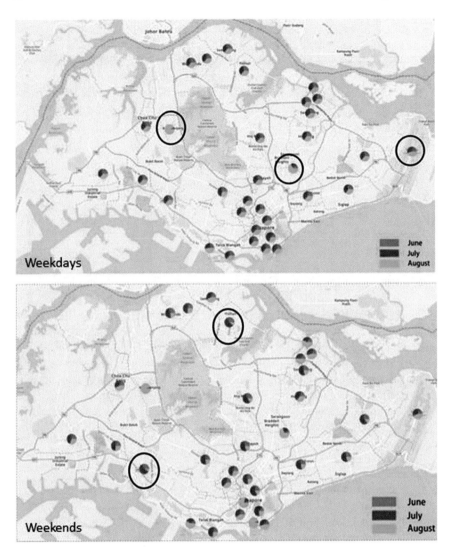

Fig. 9 Pie charts for weekdays and weekend at various popular stations

outdoor activities after Phase Two reopening. The median illustrates higher increase in the variation in popularity of commercial areas and open spaces during weekends from June to August. Interquartile range shows that the increase of popularity of open spaces varies most according to different MRT stations. Its outlier is also distinct compared to other typologies. Possible reason could be the increasing outdoor activities of working class on weekends. The reopening of some well-designed open spaces attracts more visitors for refreshment and recreation.

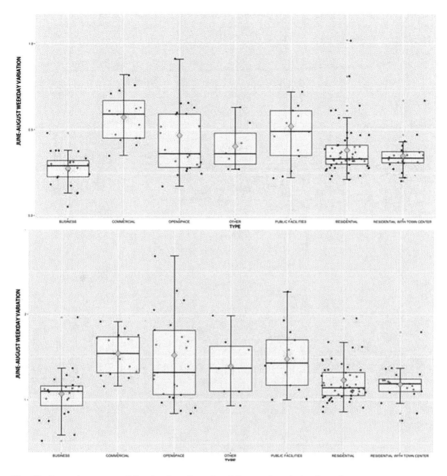

Fig. 10 June–August weekday (top) and weekend (bottom) variation boxplots

3.4 Outlier Analysis

Rank the MRT stations by JUNE–AUGUST WEEKDAY VARIATION (Fig. 11). The spikes in for JUNE–AUGUST WEEKENDs indicate the stations had higher passenger volume variation in weekend than the predicted value by weekday data.

The spikes in the difference value of WEEKDAY-WEEKEND VARIATION in June and August, indicate that there is much unnecessary travel happened in the weekend than the predicted value by weekday data. The spikes picked from both two figures are considered as the popular MRT stations.

Based on the results, eight popular stations, namely Bayfront, Labrador Park, Marina Bay, Marina South Pier, Harbour-front, Botanic Gardens, Changi Airport and Kranji stood out as being the most popular.

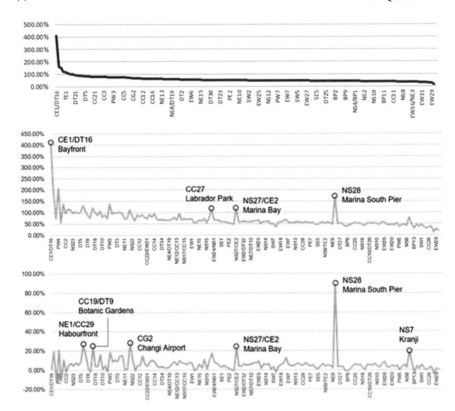

Fig. 11 June–August patterns. (Top – weekday, Mid – weekend, Bot – Weekday-weekend differential)

3.4.1 Pattern-Based Classification

Based on the WEEKDAY MEAN PV and WEEKEND MEAN PV from June to August, MRT stations could be further classified into three categories (Fig. 12) as below:

- WEEKEND MEAN PV grew rapidly from June to July, and became stabled from July to August: Changi Airport, Habourfront and Botanic Gardens;
- WEEKEND MEAN PV kept rapidly growth from June to August: Bay Front and Mirna South Pier;
- WEEKEND MEAN PV lower than WEEKDAY MEAN PV: Kranji, Marina Bay and Labrador Park.

Sifting out the third type of MRT stations that less popular in weekend, the following part will discuss what makes the first two type of MRT stations popular in the post-pandemic.

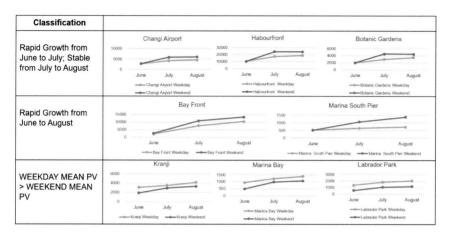

Fig. 12 Classification of popular MRT stations

4 Discussion

4.1 Open Space and MRT Station Popularity

Except for Changi Airport, the remaining four stations belonged to open space typology has high passenger volume. This is indicative of open space being more attractive to people post easing of Covid-19 restrictions. The other popular places were Harbour front, Bayfront and Marina South Pier which were located along the waterfront and presented a combination of leisure and commercial activities.

The rapid passenger volume growth from June to August helped identify the MRT stations which had the most reduced passenger volume during the Circuit Breaker. The two stations, Bayfront and Marina South Pier, which do not have residential land use within the 700 m radius buffer were particularly the ones most impacted. Residential land use associated MRT stations had a stable passenger volume irrespective of the restricted phases. MRT stations which supported commercial and business typologies were the ones which fluctuated the most before and after the circuit breaker. The gradually growing passenger volume from Phase One onwards indicated the strong residence brought by open space and attractive activities.

4.2 Re-opening Policy and Passenger Volume

All the stations responded to the reopening policy in June. The temporal changes to passenger volume are summarized in Table 1. Especially the phase two of re-reopening from 19 June. Recreation and activities gradually reopened, and park facilities including playgrounds, beaches, carparks etc. also reopened to the public.

Table 1 Passenger volume change summary

	Rapid growth from June to July, stable from July to August			Rapid growth from June to August	
	Chanqi Airport	Harbourfront	Botanic Gardens	Bayfront	Marina South Pier
Pre	22 March: Singapore banned all short-term visiting from entering or transiting May: Close of Terminal 2 and 4		10 April: Closed	22 March: Close of Bayfront Park	
June	2 June: Gradually reopen transit 19 June: Reopen Jewel Changi Airport providing food and retail	19 June: Reopen certain offerings in Sentosa (attractions remain closed)	19 June: Reopen	19 June: Marina Bay Sands began first stages of reopening 22 June: Reopen Bayfront Park with restrictions	19 June: reopened beaches at St John's, Lazarus & Kusu Island
July		1 July: Reopen Sentosa		1 July: Marina Bay Sands started gaming operations 17 July: Marina Bay Sands reopened along with entertainment activities	
August				1 August: 2 hotel towers reopened	

Therefore, those MRT stations with open space became popular. For Changi Airport, the close of two terminals have a great impact on its passenger volume, the reopening of Jewel Changi Airport as well as the reinstatement of flights attracted more people come to this station.

Different from other stations, Bayfront MRT station had continuous re-opening steps from June to August, stimulating the commercial and leisure activities. It is also the possible reason for its continuous growth of passenger volume. The other station with rapid growth from June to August, Marina South Pier, has one of the ferry terminals in Singapore. Citizens had increasingly willingness to enjoy the leisure activity at southern islands after the re-opening, facilitating the revitalization of Marina South Pier.

4.3 Conclusions and Limitations

Big Data can provide timely information that is conductive to interdisciplinary research and can inform decision makers with emerging pattern to make prompt respond.

Station typologies indicate the urban functions served by the station, depicting its user group and their travel behaviour.

The stations popularities in the post-pandemic reflect the travel behaviour change caused by the pandemic and related policies, suggest how some urban spaces in cities help people recovery from the pandemic.

In general, Commercial and Business type stations have higher passenger volume on weekdays, presumably because these types of stations aggregate more jobs and are daily commuting destinations. At the same time, stations with town/regional commercial centers had significantly higher passenger volume than those without, suggesting that integrated development of commercial & office space with MRT stations could improve their efficiency. Stations under the Commercial and Open space typologies are more popular on weekends, reflecting the fact that more shopping and leisure behaviors occur on weekends. Stations under these typologies need flexible design and management to cater for the denser weekend users.

With the gradual re-opening in the post-pandemic, people's travel behavior has increased. On weekdays Commercial and Public facilities stations saw significantly higher passenger volume, likely related to the return to work of nearby workplaces. On the weekends, each of these types of stations recorded higher trips than on weekdays, reflecting people's great enthusiasm for going out on weekends after the circuit breaker quarantine. Commercial, Open space, and Public facilities were the most popular types. Open space has many significant outliers, such as Botanic Gardens, Bayfront near the Gardens by the Bay, and Marina South Pier with boat access to the southern islands. Changes in the popularity of these stations reveals the importance of open space to health and well-being in the post-pandemic. However, the surge in passenger volume at these stations poses a higher risk of infection, and managers should respond quickly to this change by implementing stricter disease control measures.

4.3.1 Limitations

The project has some scope of limitations, both in terms of data acquisition and data analysis. For instance, due to the limited time range of accessible origin–destination (OD) data, pre-pandemic information was not included in this study. The land use data used for the typology extraction was based on the 2019 Singapore Master Plan. This resulted in the non-inclusion of MRT stations built after 2019. Also, there could be other unaccounted micro-factors which affect the passenger volume in Singapore which was not account in this study.

4.3.2 Future Scope

The study demonstrates the use of station passenger volume vis-à-vis land use and relaxation of COVID-19 restrictions to picture travel behaviour change. The study contributes to insights on travel patterns in Singapore which can be used to plan the frequency of MRT along sectors in response to easing of Covid-19 rules which could be a means for optimise resources and improving sustainability of transportation services. In conclusion, the study expounds on the potential of Big Data in urban planning and city management, foreseeably the Spatial Big Data can help cities to respond to emergencies more time-effectively in the future.

References

1. Berawi MA, Saroji G, Iskandar FA, Ibrahim BE, Miraj P, Sari M (2020) Optimizing land use allocation of transit-oriented development (TOD) to generate maximum ridership. Sustainability (Basel Switzerland) 12(9):3798. https://doi.org/10.3390/su12093798
2. Chakraborty A, Mishra S (2013) Land use and transit ridership connections: Implications for state-level planning agencies. Land Use Policy 30(1):458–469. https://doi.org/10.1016/j.landus epol.2012.04.017
3. Kim M, Kim S, Heo J, Sohn H (2017) Ridership patterns at subway stations of seoul capital area and characteristics of station influence area. KSCE J Civ Eng 21(3):964–975. https://doi.org/10.1007/s12205-016-1099-8
4. LTA (2013). Land transport master plan 2013[R]. Land Transport Authority, Singapore
5. Moslem S, Campisi T, Szmelter-Jarosz A, Duleba S, Nahiduzzaman KM, Tesoriere G (2020) Best-Worst method for modelling mobility choice after COVID-19: evidence from Italy. Sustainability (Basel Switzerland) 12(17):6824. https://doi.org/10.3390/su12176824
6. Niu S, Hu A, Shen Z, Lau SSY, Gan X (2019) Study on land use characteristics of rail transit tod sites in new towns—taking Singapore as an example. J Asian Arch Build Eng 5:1–12
7. Sidek MFJ, Othman NNAN, Hamsa AAK, Noor NM, Ibrahim M (2017) Summary on the effect of density, diversity and pedestrian infrastructure on the use of rail-based urban public transport. WIT Trans Ecol Environ 210:567. https://doi.org/10.2495/SDP160471
8. Zhu X, Liu S (2004) Analysis of the impact of the MRT system on accessibility in singapore using an integrated GIS tool. J Transp Geogr 12(2):89–101

A Geospatial Analysis of Tweets During Post-circuit Breaker in Singapore

Xu Yuting, Lim Zhu An, Sherie Loh Wei, and Phang Yong Xin

Abstract Since 19 July 2020, Singapore entered Phase 2 of re-opening after one and half month's "Circuit Breaker" measures to curb the spread of COVID 19. Although most businesses and public places have resumed operation at a reduced capacity, individuals were strongly advised to practice social distancing and avoid crowds. Both implicit and explicit measures to prevent overcrowding had impacted on how people visit places in Singapore. The current study used geotagged Twitter data between September to October in 2020 to examine the spatial and temporal patterns of residents' locations in Singapore and explored the service amenities which remain "attractive" to residents. Random Forest Supervised Machine Learning Model was used to train and predict spatial distribution of activities during off-work recreational hours using service amenities point of interests (POIs) and land use merge. Five explanatory variables used were parks, public links between parks and malls, taxi stands, residential areas, and shopping malls which had the strongest influence in driving the model prediction of spatial distribution of activities in off-work recreational hours. While distinct temporal patterns of tweets were expected during office hour, this analysis revealed no such statistically significant clusters. The regression analysis showed that distances to service amenities did not provide strong explanations for tweeting patterns.

X. Yuting (✉)
45 Hindhede Walk, #03-04, Singapore 587978, Singapore

L. Zhu An
501 Dunman Road #11-02, Singapore 439193, Singapore

S. Loh Wei
661 Hougang Avenue 4 #11-381, Singapore 530661, Singapore

P. Yong Xin
Blk 94 Bedok North Avenue 4 #05-1385, Singapore 461094, Singapore

© The Author(s), under exclusive license to Springer Nature Singapore Pte Ltd. 2022   79
S. N. Kundu (ed.), *Geospatial Data Analytics and Urban Applications*,
Advances in 21st Century Human Settlements,
https://doi.org/10.1007/978-981-16-7649-9_5

1 Introduction

The ongoing COVID-19 pandemic has severely impacted the Singapore economy and society. On 7 April 2020, the Circuit-Breaker (CB) measures were implemented to restrict the usage of public spaces as well as social gathering [3]. The Circuit Breaker was subsequently lifted on 2 June [4] and the country entered into a relaxed but cautious phases of re-opening. In late October 2020, small group gatherings were allowed amid re-opening of most businesses except for high-risk ones such as bars, pubs and karaoke services [5]. The steps reflected a careful balance of the economic interest against the backdrop of the risk of virus transmission. Individuals were advised to practice social distancing, while businesses were advised to reduce service capacity to avoid crowding. Government agencies rolled out several apps such as URA's Space Out and NParks' Safe Distancing @ Parks to inform residents on the crowd level at public places, for people to make informed decisions on when and where to go for needed services. Besides the implicit consideration of visiting a public pace more explicit movement control measures, such as zoning and no-visitor rules, was also implemented by organisations such as NUS [2]. At the same time, more Singaporeans have opted for having services delivered to where they live rather than them moving out of homes to fetch them [17]. The unique context post-CB generates interest to understand what activity patterns are like currently, and whether services and amenities still attract people to visit them.

Location-based social media data, contributed by individual users voluntarily, offers a wealth of information, including temporal, spatial and semantic attributes, and has been used in a growing body of literature [19]. In domains of urban science and GIS, social media data has been used to understand individual activity-space and spatiotemporal public space visiting patterns [9], sentiment of the public near places [9, 18], or crowd behaviours during a major event [10]. Soliman et al. [15] explored urban land use classification using movement patterns of Twitter users in Chicago, achieving an accuracy of 0.78 with key locations derived from temporal patterns. In a similar vein, other social media platforms like Foursquare that have check-in features have been employed in the study of land use in New York City, in addition to data extracted from Twitter [20]. They then adopted a clustering inference approach in tandem with a supervised learning approach (random forest classifier) for land use inference, where the latter method returned a higher overall accuracy of approximately 79%. Tweets have been analysed to assess public sentiments and mobility patterns during the novel coronavirus pandemic [8] in the USA. The findings established a positive correlation between the volume of movement and infection rates.

In Singapore, however, most studies using Twitter data focused on analysing semantic information, and very few used Tweets to understand urban places. Prasetyo et al. [12] found an association between the location of schools, type and competitiveness and a connection between these educational institutions and shopping malls. There is also a gap in knowledge on how the unique context of post-Circuit Breaker re-opening has changed place-visiting patterns, and whether Twitter data was suitable for studying Singapore's urban spaces. In this study, we used Twitter data to explore the spatial–temporal patterns of activities in post-CB Singapore. Based on the place-visiting patterns established, we then modelled and predicted visitors at places during non-office recreational hours and examined how such patterns relate to service amenities. As an exploratory exercise to acquire an overview of spatio-temporal tweet patterns, the findings will also enable a discussion on the suitability of Twitter data to understand geospatial questions in the context of Singapore.

2 Methodology

2.1 Data Collection

2.1.1 Tweets

The fundamental data used for this study was tweets. The twitter public standard API v1.1 (https://developer.twitter.com/en/docs/twitter-api/v1/) was used to search for all tweets posted from 18 September 2020 to 10 October 2020. A spatial extent of 25 km radius from the Central Catchment was defined for tweet collection via the API, an area that effectively covers the whole of the Singapore (Fig. 1).

Twitter API returned each tweet object with relevant attributes, that included user details, full text of tweets, timestamp and tweet location (Table 1). Coordinates and Place were the attributes which included spatial information. Coordinates were rarely included data and therefore information provided in Place attribute was used to decipher the location of the tweet. The Place attribute offered varied levels of spatial precision (Table 2) depending on the account setting of each user.

According to Twitter policy, users who chose to enable precise location on their account will have all their tweets geotagged automatically to specific coordinates, while users who chose not to enable precise location may still optionally tag individual tweets to a location by keying in a place name. Tweets with exact X/Y coordinates were digitised into point features directly, while tweets with a specific landmark as place name are geocoded using OneMap search API (https://docs.onemap.sg/#search) by providing the Place_Name attribute as keywords to obtain the coordinates. Because some place names in the tweets are erroneous or not recognised in the OneMap API, only 306 tweets with specific landmark place names were successfully geocoded. Tweets with spatial information at the scale of a neighbourhood and above were recognised to give a poor indication of the user device location and were not used in spatial analysis.

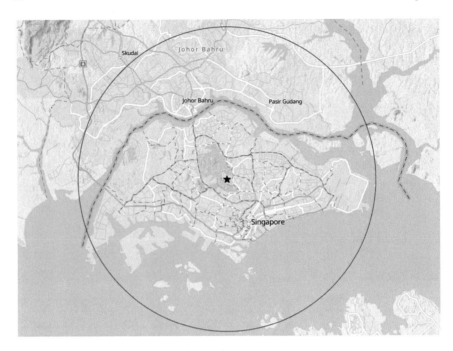

Fig. 1 Spatial Extent Tweet search and extraction

Table 1 Attribute Returned from Twitter Standard Search API

Attribute	Data type	Description
Created_at	Time	UTC Time of Tweet
Id	String	Unique ID of Tweet
Truncated	Boolean	Indicated whether the tweet was truncated or not
Text	String	Content of the tweet
Source	String	The source application used to post the tweet
URL	String	The URL through which the tweet was posted
User	String	Name, Screen name and Location of the user from twitter profile
Coordinates	Point	Latitude and longitude of the location of the tweet
Place	String	A unique id in twitter database, URL, Place Name, Country, bounding box etc
Language	String	Language of the tweet

2.1.2 Spatial Data

Geospatial locations of service amenity [11] point of interests (POIs) were collected from various official data sources and open data sources. These, listed in Table 3, were mostly available is GIS format ready for spatial analysis.

Table 2 Spatial attributes in Tweets

Type of Spatial Attributes (in decreasing order of precision)	Percentage of Tweets (%)
With exact coordinates (X/Y)	0.25
With place name (a specific landmark), but without X/Y coordinates	0.31
With place name (a neighbourhood/town), but without X/Y coordinates	0.01
With place name (a region), but without X/Y coordinates	2.43
With place name (a city), but without X/Y coordinates	0.03
With place name (a state/province), but without X/Y coordinates	0.00
With place name (a country), but without X/Y coordinates	0.05
No explicit spatial information	96.91

Table 3 Spatial data and sources

Data	Source
Hawker centres Residential with 1st storey commercial Community clubs Parks Park connector network Dual use scheme (DUS) sports facilities SCDP park mall Master plan 2019	data.gov.sg
Shopping malls	List of shopping malls in Singapore from Wikipedia (https://en.wikipedia.org/wiki/List_of_shopping_m alls_in_Singapore) and geocoded using OneMap search API
MRT stations Bus stops Taxi stands	mytransport.sg
Medical facilities	KML file extracted from Google Maps

2.2 Spatio-Temporal Analysis

Digitised tweet point features were analysed for their spatial and temporal patterns. The framework used by [13] was adapted for this study (Table 4). Building on the assumption that presence in space, proxied by posting of tweets, was driven by the purpose to receive services, Ordinary Least Squares (OLS) regression was performed on the data to estimate the relationship between aggregated tweet counts and the distance to the nearest service amenity POI. The OLS regression model was first run with distances to all 12 types of service amenity POIs as explanatory variables and the model accuracy was assessed. Explanatory variables that had Variance Inflation

Table 4 Categories of spatiotemporal analysis *adapted from* [13]

	Temporal analysis	Spatial analysis	Dynamic spatiotemporal analysis	Static spatiotemporal analysis
Temporal attribute	Independent	Fixed	Independent	Fixed
Spatial attribute	Fixed	Independent	Dependent	Dependent
Thematic attribute	Dependent	Dependent	Fixed	Fixed
Examples of questions for this study	In the same location, how do activity patterns vary with time?	In the same time period, how do activity patterns vary in places?	How do spatial patterns of activities vary with time?	How do spatial patterns vary in different locations at a fixed point of time?

Factors (VIFs) > 7.5 and collinearity > 0.6 were excluded from the model. The final OLS regression model was then again analysed for its accuracy and applicability.

2.3   Random Forest Model

A random forest regression model was chosen to model the number of tweets at different locations across the study area. The forest-based classification and regression tool in ArcGIS was an adaption of Leo Breiman's random forest algorithm, which is a supervised machine learning model. This suited well for the purposes of this study. With the input of service amenity POIs as point/polygon feature layers, the tool was able to generate distances from each feature to be used for the training of the regression model. Presence of land uses and distance to amenities were conceptualized as potential explanatory variables to model and predict spatial distribution of people in the area.

2.3.1   Model Inputs

A square fishnet of 1 sq km units were used to tag the attributes of each tweet point feature for overlay analysis. Therefore, the whole area of Singapore was divided into one sq km grid cells. At this resolution, a good balance between the model and GIS processing speeds were achieved. It also potentially increased the number of land use types per fishnet grid. The presence of land use (based on Master Plan 2019) in each grid was denoted with Boolean attributes (1 for presence, 0 for absence). Tweets posted on Weekends as well as Weekdays from 6 pm to midnight were aggregated into each grid cell to produce a total count that represented spatial distribution of people during off-work recreational hours. A total of 3982 tweets were used in this model input. The land use and tweet count attributes were then attached to the fishnet centroid point features for modeling.

Besides land uses, service amenity POIs (Table 3) was used as explanatory training distance features. The tool generates the distance between each training feature point and all the distance variable features, which will then be used as continuous numerical inputs for the training of the model. Attributes from the different fields within the training feature layer are used as variables for the model as well.

2.3.2 Model Settings

The model was initially trained with 100% of the training data to first determine the key explanatory variables that were driving the results of the model. The model parameter "Number of trees" was modified to determine a stable model in terms of the ranking of importance variables. After a few iterations, the number of trees chosen for this model was reduced to 500 trees. The other model parameters for the random forest classifier, like minimum leaf size, maximum tree depth, data per tree, were all based on the default settings.

2.3.3 Model Validation

After obtaining a stable model with training data, the model was validated by splitting the training data into smaller parts. This was necessary for determining the accuracy of the model. 10% of the training data was used to validate the accuracy of the model. The model was trained without this random subset of data, and the observed values for those features were compared to the predicted values to validate model performance. Because the subset data used for validation was random, a total of 15 validation runs were performed to obtain a maximum, minimum, and median R-Squared value for the accuracy validation.

3 Analysis and Results

3.1 Twitter Users in Singapore

As per Fig. 2a1, a total of 748,533 users contributed to 2,776,651 tweets collected in the full circular buffer area from 18 September 2020 to 10 October 2020, with an average of 3.71 tweets per user. The count of tweets posted by each user varied remarkably, with one user posting over 9000 tweets, while a majority of the rest posting only 1 tweet within the study period. A similar pattern was found among users who have posted geotagged tweets within Singapore. A total of 1396 users had posted geotagged tweets on Singapore main island, among which, more than half

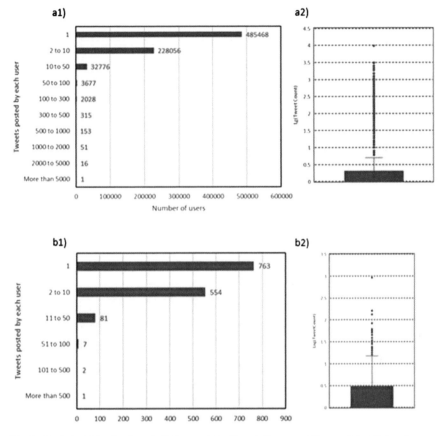

Fig. 2 Statistical Distribution of all Tweets (**a1**, **a2**) and Geotagged Tweets (**b1**, **b2**) (total tweets = 2,776,651, total user = 748,533, mean tweets/user = 3.71, median tweets/user = 1, sd = 25.46), **b1** Distribution of geotagged tweets in Singapore (total tweet count = 5,754, total user = 1,396, mean tweets/user = 4.12, median tweets/user = 1, sd = 25.87)

of the users posted only 1 tweet within the study period. The figures suggest that only a small handful of users were contributing to the greatest number of tweets. This finding was consistent with several other studies which used Twitter data in Singapore [12] and was also consistent with studies in other countries [1].

3.2 Distribution in Space and Time

Daily total count of tweets and geotagged tweets in Singapore shows a highly variable pattern over the data collection period (Fig. 3). Days with lower-than-average tweets and geotagged tweets generally corresponded with missing periods of tweet collection.

Two peak periods for tweet activities on weekdays from Fig. 4: during morning commuting hours (6–8 AM) and evening after-office hours (after 6 PM). This could be due to availability of users to engage in online activities. On weekends, the hourly plot shows a smoothly increasing trend for average tweet counts, suggesting that on weekends, online activities grow in intensity gradually over the course of a day, and finally reaching the peak in the late evening. Over the course of a week, fewer tweets are posted during office hours (weekday 9AM–6PM) than other time periods.

Results from Incremental Spatial Autocorrelation on the entire Singapore geotagged tweets dataset showed peak distance at around 1 km. This was the spatial scale at which clustering could be observed, but a low z-score suggests that such clustering is not statistically significant. Tweets were further segmented into 2-h intervals and Incremental Spatial Autocorrelation was conducted on each time block. For all time blocks, the result showed no statistically significant spatial clusters. This is suggestive of the fact that processes which promote global spatial clustering are likely to be random (Fig. 4).

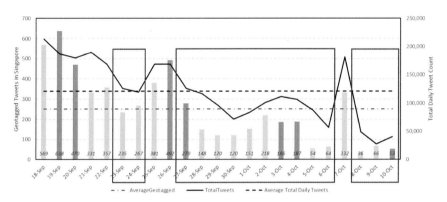

Fig. 3 Total daily count of tweets and geotagged tweets in Singapore collected over the study period

Hour of Day	0	1	2	3	4	5	6	7	8	9	10	11	12	13	14	15	16	17	18	19	20	21	22	23
Monday	11.33	7.00	4.00	4.33	2.00	3.33	13.00	12.67	3.67	4.33	5.33	5.00	4.67	3.67	6.00	10.67	13.00	12.00	12.33	14.33	31.00	21.67	16.67	28.00
Tuesday	17.33	6.33	7.67	4.33	3.33	6.00	15.67	24.67	7.00	5.67	4.00	7.67	5.33	9.67	6.00	3.00	5.33	5.00	6.00	12.33	14.00	18.33	10.67	10.00
Wednesday	16.67	5.67	4.33	2.00	4.33	3.00	13.67	11.67	0.00	0.00	0.33	5.33	7.33	8.00	5.00	10.33	7.67	23.67	16.33	15.33	15.00	21.33	18.33	25.33
Thursday	12.67	6.67	4.33	3.33	3.00	2.67	11.33	14.67	3.33	0.00	0.00	0.00	0.00	0.00	0.00	0.00	2.67	10.00	17.33	11.00	19.00	19.67	13.67	13.67
Friday	5.75	3.75	2.50	3.00	2.50	1.00	11.00	6.50	10.00	8.00	12.25	9.75	11.75	14.75	13.00	12.50	15.25	12.50	23.00	28.25	27.75	24.50	23.50	25.75
Saturday	13.25	5.25	4.50	5.00	2.00	2.75	5.75	7.25	8.25	12.75	19.50	17.50	15.00	14.50	17.00	14.25	13.75	13.75	25.25	26.75	31.50	25.50	25.00	16.25
Sunday	11.00	5.33	6.33	3.00	2.00	4.33	9.67	7.33	8.33	5.33	13.00	7.33	10.00	6.67	8.00	8.67	19.67	16.33	25.00	18.33	19.00	30.00		21.67

Fig. 4 Average hourly count of tweets posted within Singapore main island spatial extent

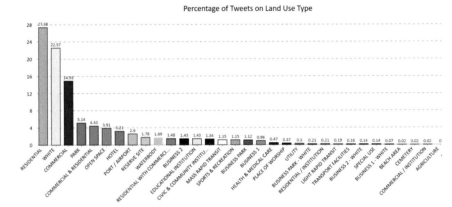

Fig. 5 Breakdown of land use types with tweet counts over the study period

Kernel density of tweet point features (Fig. 6) showed some areas having higher density of tweets, such as the Central Business District (CBD), and several large residential estates and regional hubs e.g. Choa Chu Kang, Bukit Batok, Woodlands, Hougang-Sengkang-Punggol stretch, Serangoon, Bishan-Ang Mo Kio, Changi Airport. Tweet counts were then summed up based on the land use types from Singapore's Master Plan. It was found that over 27% of tweets were posted from Residential land plots, followed by White and Commercial land use types (Fig. 5), possibly because work-from-home was still the default mode of working for most businesses and organisations during the data collection period.

The central area presents higher density of tweets than other areas of Singapore in all time periods, suggesting a consistently higher intensity of tweet activities than other areas in Singapore. The observation was also made by [12] on tweets in 2012. This can be possibly explained by a greater diversity of land use mix and more important services in the central area, that may attract both work and leisure purpose-driven trips.

Over the course of a day, time period between the 04:00 - 06:00 h represented the block with the lowest spatial coverage of tweet activities, forming distinct islands of tweet density in the CBD and big residential estates such as Bukit Batok, Serangoon, Bedok and Seng Kang. Night-time tweets cover a much bigger and contiguous geographical area, suggesting more distributed patterns of tweeting activities in off-work hours.

From the spatiotemporal analysis above we conclude that tweeting patterns exhibit distinct diurnal and weekly trends. Although visually we can identify distinct areas with higher density, these areas do not form statistically significant spatial clusters, suggesting that random processes could have contributed to the formation of such high-density areas in post-CB Singapore.

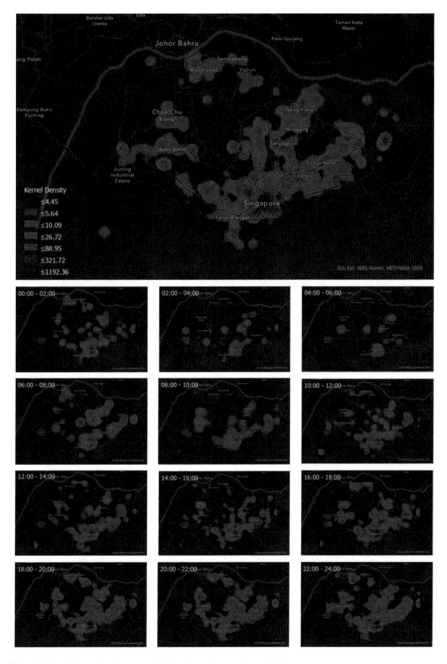

Fig. 6 Kernel density of tweets during the entire study period

3.3 Regression Analysis

Among the 12 explanatory variables in the OLS regression model, distance to Community Centres was removed for reflecting a high VIF value, and MRT, Hawker Centres, Bus Stops and Medical Facilities were removed for high correlation R-Squared values. The final OLS regression model used the remaining seven explanatory variables: Taxi Stands, Shopping Malls, Residential with 1st Storey Commercial, Parks, Dual Use Scheme (DUS) sports facilities, SCDP Park Malls, and Park Connector Network.

Results from the revised model (Tables 7 and 8) show that it has improved substantially from the first model (Tables 5 and 6). VIF values for the remaining variables are lower, suggesting that the remaining variables exhibit lower redundancy and multicollinearity, and are suitable explanatory variables. The significance of the Koenker (BP) Statistic has increased, as shown in a reduced probability from 0.95 to 0.71. Hence, we can conclude that the 7 variables surfaced in Fig. 11 have the greatest influence on the number of tweets recorded.

To illustrate further (with reference to Table 6), the relationship between "NEAR_DUS" and the dependent variable is a positive one, where for every 1 m increase in distance from this facility, the tweet count increased by 0.000877. On the contrary, there is an inverse relationship between "NEAR_PARKS" and tweet count - with a 1 m increase in distance from a park, the number of tweets fall by 0.001257. Taking the absolute number from the coefficients, the influence of the 7 remaining variables were ranked (Table 9).

Despite showing improved variable suitability, the final model remained weak in performance, as seen in a low Multiple R-Squared and Adjusted R-Squared value (Table 8), suggesting that the model is able to account for a very low percentage of dependent variables. This could be because the model is missing key explanatory variables or that a linear regression may not be a suitable model choice. The low coefficient values for all explanatory variables suggest that the relationships between all of them with the dependent variable (tweet count) are weakly positive or negative. It should also be noted that none of the explanatory variables are associated with a statistically significant coefficient, and a significant probability of Jarque–Bera Statistic suggests that the model is highly biased. In Fig. 7, it was observed that the residueals deviate from the line of the best fit and are thus not normally distributed. This goes against one of the classical assumptions of OLS, providing estimates that were biased with high variances. Additionally, the skewness (normally distributed $= 0$ vs. dataset $= 47$) and kurtosis (normally distributed $= 3$ vs. dataset $= 2266$) values are far from that of a typical normally distributed dataset.

In this view, non-parametric tests that are free from any underlying assumptions of the dataset distribution could be more suitable for this particular or adjacent studies (e.g. Mann Whitney U Test, Kruskal Wallis H Test). Adopting a log-linear scale in the model may improve the estimates and their significance as well.

Table 5 OLS results for all 12 explanatory variables in the first OLS model (the asterisk denotes a statistically significant explanatory variable)

Variable	Coefficient [a]	StdError	t-Statistic	Probability [b]	Robust_SE	Robust_t	Robust_Pr [b]	VIF [c]
Intercept	2.617869	0.977584	2.677898	0.007457*	0.51509	5.082355	0.000001*	
N EAR_TS	0.000327	0.001376	0.237558	0.812246	0.000311	1.050645	0.29352	5.077168
N EAR_SM	−0.00012	0.001023	−0.11284	0.910151	0.000161	−0.71629	0.473876	3.844214
NEAR_R1C	0.000402	0.000841	0.477476	0.633082	0.000292	1.375138	0.169232	3.966233
NEAR_PARKS	−0.00167	0.00104	−1.60086	0.109557	0.001037	−1.60524	0.108589	4.19625
NEAR_MRT	−0.00099	0.001275	−0.77247	0.439901	0.000967	−1.01871	0.308433	5.082725
NEAR_MF	−0.00013	0.000756	−0.17459	0.861411	0.000197	−0.67134	0.502068	4.615759
NEAR_HC	0.000095	0.001034	0.091984	0.926702	0.000152	0.62578	0.531521	5.683691
NEAR_DUS	0.000359	0.00083	0.432495	0.665437	0.000784	0.45738	0.647455	4.034707
NEAR_CC	0.001166	0.001413	0.825607	0.409097	0.000638	1.828182	0.067651	10.58637
NEAR_BUS	0.001095	0.002416	0.453044	0.650574	0.000963	1.136671	0.255788	4.423079
NEAR_SCDP	0.00003	0.000611	0.04914	0.960798	0.000112	0.266909	0.789568	3.002643
NEAR_PCN	−0.0003	0.000688	−0.43547	0.663274	0.000348	−0.85963	0.390064	2.571266

Table 6 OLS diagnostics for all 12 explanatory variables (the asterisk denotes a statistically significant explanatory variable)

Input features	tweetsaggregated2	Dependent variable	ICOUNT
Number of observations	2369	Akaike's Information Criterion (AICc) [d]	21,538.41032
Multiple R-Squared [d]	0.002627	Adjusted R-Squared [d]	−0.002453
Joint F-Statistic [e]	0.517121	Prob(>F), (12,2356) degrees of freedom	0.905105
Joint Wald Statistic [e]	31.356328	Prob(>chi-squared), (12) degrees of freedom	0.001738*
Koenker (BP) Statistic [f]	5.34448	Prob(>chi-squared), (12) degrees of freedom	0.945473
Jarque–Bera Statistic [g]	502,292,827	Prob(>chi-squared), (2) degrees of freedom	0.000000*

3.4 Modelling Using Random Forest Classifier

The summary of the regression diagnostics results is shown in Table 10. R-Squared value of the validation data suggests that the model can explain 50.7% of the observed variation, and it is statistically significant (i.e. P-value < 0.025). This indicated that the model might still lack key explanatory variables, which will improve the performance of the model. The uncertainty in the input data due to the nature of Volunteered Geographic Information (VGI) data also affects the performance and accuracy of the random forest model.

The top 20 variables are shown in Table 11. These variables were ranked according to their importance in driving the results of the random forest model. From the results of the top variable importance ranking, it is identified that the location of parks has the most significant influence in terms of driving the model outcome when training with the total tweet count data per 1 km cell. It was also noted that the top 5 variables of Parks, SDCP Park and mall public link, Taxi stands, Residential, and Shopping malls makes up more than half of the variable importance. This indicates that these 5 variables are highly influential in training of the random forest model.

4 Discussion

4.1 Activity Patterns and Relations to Service Amenities and Land Use

In this study, we used tweet activity locations to gauge physical activity patterns in urban places. Distinct diurnal and weekly activity patterns that closely resemble a typical office work temporal profile was observed. Generally, more twitting activities

Table 7 OLS results for final model variables in the revised OLS model

Variable	Coefficient [a]	StdError	t-Statistic	Probability [b]	Robust_SE	Robust_t	Robust_Pr [b]	VIF [c]
Intercept	2.177624	0.774149	2.812927	0.004953*	0.227043	9.591237	0.000000*	
N EAR_TS	0.000199	0.001102	0.180254	0.856961	0.000178	1.117468	0.263903	3.258452
N EAR_SM	−0.0003	0.000861	−0.34902	0.727122	0.000312	−0.96202	0.336122	2.728471
NEAR_R1C	0.000613	0.000751	0.816322	0.414385	0.00033	1.857277	0.063396	3.166298
NEAR_PARKS	−0.00126	0.000921	−1.36567	0.172183	0.000857	−1.46672	0.142601	3.293348
NEAR_DUS	0.000877	0.000621	1.412313	0.158003	0.001017	0.861559	0.389003	2.262284
NEAR_SCDP	0.000045	0.000592	0.07605	0.939369	0.000105	0.42848	0.668356	2.823583
NEAR_PCN	−0.00027	0.000647	−0.41938	0.674995	0.000375	−0.7232	0.469622	2.281214

Asterisk denotes a statistically significant explanatory variable

Table 8 OLS diagnostics for final model variables in the revised model

Input features	tweetsaggregated2	Dependent variable	ICOUNT
Number of observations	2369	Akaike's Information Criterion (AICc) [d]	21,529.78831
Multiple R-squared [d]	0.002004	Adjusted R-Squared [d]	−0.000955
Joint F-statistic [e]	0.677156	Prob(>F), (7,2361) degrees of freedom	0.578026
Joint wald statistic [e]	17.811667	Prob(>chi-squared), (7) degrees of freedom	0.012849*
Koenker (BP) statistic [f]	4.568018	Prob(>chi-squared), (7) degrees of freedom	0.712512
Jarque–Bera statistic [g]	502,681,626.5	Prob(>chi-squared), (2) degrees of freedom	0.000000*

Asterisk denotes a statistically significant explanatory variable

Table 9 Ranking of variables based on OLS

Ranking	Variable	Coefficient (Absolute)
1	Park (NEAR_PARKS)	0.001257
2	Dual Use Scheme (NEAR_DUS)	0.000877
3	Residential with 1st Storey Commercial (NEAR_R1C)	0.000613
4	Shopping Mall (NEAR_SM)	0.000301
5	Park Connector Network (NEAR_PCN)	0.000271
6	Taxi Stand (NEAR_TS)	0.000199
7	SCDP Park Malls Public Link (NEAR_SCDP)	0.000045

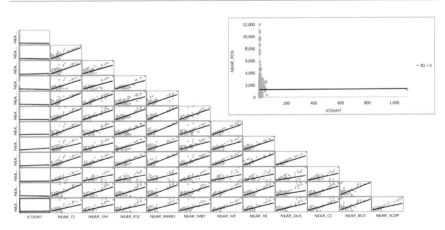

Fig. 7 Variable residuals

Table 10 Summary of random-forest model validation results

	Training data	Validation data
R-squared	0.868	0.507
P-value	0.000	0.000
Standard error	0.005	0.086

Table 11 List of explanatory variables ranked by importance

Rank	Variable	Importance	Percentage (%)
1	PARKS	157,902.84	22
2	SDCP PARK MALLS PUBLIC LINK	104,611.88	14
3	TAXI STANDS	75,827.32	10
4	RESIDENTIAL	72,528.73	10
5	SHOPPING MALLS	54,777.42	8
6	RESIDENTIAL WITH COMMERCIAL AT 1ST STOREY (Continuous, Distance Feature)	48,148.46	7
7	SCHOOLS	30,755.03	4
8	HOTEL	30,067.48	4
9	MEDICAL FACILITIES	24,600.19	3
10	MRT STATIONS	21,550.71	3
11	COMMUNITY CLUBS	21,163.32	3
12	PARK CONNECTORS	14,511.27	2
13	HAWKER	13,645.71	2
14	COMMERCIAL and RESIDENTIAL	13,147.71	2
15	RESIDENTIAL WITH COMMERCIAL AT 1ST STOREY (Categorical, Land use type)	9161.57	1
16	SPORTS and RECREATION	8389.23	1
17	COMMERCIAL	6670.87	1
18	BUS STOPS	6664.18	1
19	PLACE OF WORSHIP	4576.85	1
20	EDUCATIONAL INSTITUTION	3058.23	0

occurred beyond office hours, and weekend activity increased gradually as the day progressed and no distinct peak activity periods were observed. This could mean that majority of Twitter users in Singapore are likely to be engaged typically during working-hour schedules and tend to post more tweets outside of office hours.

Another important observation from the spatiotemporal analysis was that there was no statistically significant spatial cluster, despite showing areas with visually higher tweet activity density. The processes that contribute to the observed varied spatial activity density could be a result of random processes, and further segmentation is required to uncover the mechanism of such processes and their influence on activity patterns. Further studies could be conducted to assess if such patterns would support the hypothesis that the current safe distancing measures, both implicit and explicit, are effective in preventing overcrowding.

Activity patterns in post-CB Singapore could not be adequately explained by distance to service amenities. OLS regression was performed with the assumption that service amenities could attract human activities, but the result suggests poor model accuracy and absence of significant variable coefficients. Hence, distances to service amenities are not strong explanatory factors for activity patterns in post-CB Singapore. The Random Forest Model showed improved model performance by focusing on activities in recreational hours and also accounting for land use mix. The result suggests that several amenities and land use types, such as parks, shopping malls, residential developments and HDB shop houses are among the more important explanatory variables for activity patterns post-CB. It should be noted that residential land is also the top land use type where tweet activities are found, but were not included in the OLS regression model, possibly contributing to the poor OLS regression model performance. The high percentage of tweets from residential land and its relatively strong explanatory factor might be explained by the fact that work-from-home was still the default mode of working during the data collection period. Therefore, locations of residential land might have accounted for most tweet activities post-CB.

4.2 Limitations

Prior to this case study, Twitter data was rarely used to study activity patterns in Singapore and there was no evaluation of suitability of Twitter data to study urban place related issues. Several limitation can be attributes to sparse use of twitter data which can be improved if geographic information is embedded into tweets. Spatialization of tweets based on implicit geographical indicators too could have introduced some ambiguity into the cell-based aggregations and analysis.

4.2.1 Twitter Data in Spatial Context

This study shows that Twitter user group is a highly biased representation of the Singapore resident population. Only a small number of users, as compared to the resident population, posted tweets that can be geotagged to a precise location in Singapore. Due to absence of other user details, such as age, race and gender, we are unable to analyse if the Twitter user group constitutes a representative sample of the resident population. However, according to [16], a provider of market and consumer data, Twitter is the sixth most used social media in Singapore, with estimated 1.37 million users in 2020, mostly aged between 25 and 34. This suggests that the tweets collected only represent a small group of young working adults, which might support the temporal patterns that closely resemble a typical office work patterns. Among the Twitter users, each user is unevenly represented, with the majority of users only posting 1 tweet over the whole period of data collection, and a small handful of active users contributes to a large majority geotagged tweets. This means that the spatial patterns derived from geotagged tweets are contributed by a small number of active users who geotagged many of their tweets.

On the quality of spatial attributes, we found varied levels of spatial precision of tweet location. Only 0.28% of all tweets were geotagged to precise X/Y coordinates and another 0.31% were geotagged to a specific landmark, while over 90% tweets do not have any discernible location attributes. Among the 0.31% tweets that were geotagged to a specific landmark, a large majority could not be directly geocoded with OneMap API, due to erroneous entries. The remaining tweets, despite having place name attributes, are too generic to provide insight on exact location of users and activities.

Tweeting habits may influence the data quality as well. The study relies on voluntarily posted tweets to understand temporal and spatial activity patterns. It is well documented that not all places receive equal coverage on social media, as users may selectively report locations only when they are deemed important to be shared [14].

To prevent heavy API usage from disrupting its network, Twitter has imposed a limit on how many tweets can be queried with its free-of-charge public API. The number of tweets collected will not exceed 1% of all tweets posted by users. Twitter does not disclose the detailed mechanism to allow 1% of tweets to be accessed via the API. Assuming that a random sampling process is deployed by the API, the tweets collected could similarly be more representative for users who have posted more, than the infrequent users. This limitation can be addressed by using the chargeable premium API from Twitter, which allows access to the full Twitter dataset. [12] collected one month's full Tweet dataset with premium API and managed to gather 38,646 Singapore Twitter users who posted geocoded tweets over one month in 2012. Although they used user profile location instead of tweet location to filter the tweets and it was known that user profile location could differ from real-time tweet device location [6], data from premium API captures much more users and tweets and could present a viable solution for several limitations mentioned above.

Many of the above-mentioned weaknesses of Twitter data are characteristic to VGI big data due to absence of quality control on user-generated content. However, Twitter data still represents a valuable dataset that is accessible to the general public, and contains rich temporal, spatial and semantic attributes that can be further exploited with a suitable research question that adequately addresses data bias and quality issues.

4.2.2 Other Limitations

Absence of Pre-CB Activity Pattern as a Benchmark

In this study, we presented findings on activity patterns in the context of post-CB re-opening. An understanding of the pre-CB activity patterns and correlations with service amenities will allow a comparison of pre- and post-CB activity patterns and assess the effectiveness of safe distancing and risk communication measures on physical activity patterns.

Missing Time Periods in Data Collection

The missing blocks of time when tweet collection was unsuccessful pose challenges for an accurate understanding of temporal tweet patterns. In this study, we had attempted to minimise the impact by segmenting the datasets into time periods for separate analysis. However, a full and complete period of data collection would still be beneficial to present an unbiased study on the activity patterns post-CB.

Geocoding

There are three limitations to the geocoding of the tweets data that this study had to consider. Firstly, the data collected in the Place_Name field is not always valid. It may contain invalid tokens, improper place names, spelling mistakes, typographical errors, etc. This will result in an unsuccessful conversion and indicates that these records will be excluded from the geocoding process. Secondly, it was noted that there were many place names with a general location rather than a specific location name (e.g. Central Region). These place names will return a valid coordinate but are deemed as inaccurate. Lastly, some place names might occur in more than one location. For example, restaurants or supermarkets might have multiple branches throughout Singapore (e.g. Fairprice), therefore the coordinates returned might not be accurate. The geocoding pipeline in this study accounts for the limitations mentioned above.

In studying mobility behaviours using the tweets itself, the risk of skewed data remains with the attachment of fake locations that is not easily detected by GIS tools [7]. The accuracy of the geocoding process is only as accurate as the quality of the input data.

5 Conclusion

This study presents an attempt to explore spatiotemporal activity patterns and model the relationship between activity patterns with service amenities in post-CB Singapore using Twitter data. The findings offered preliminary insights into place-visiting patterns in Singapore under the explicit and implicit orders to prevent overcrowding to curb disease spread in Phase 2 of re-opening after Circuit Breaker. We showed that distances to service amenities are not strong explanatory variables for the observed activity patterns in post-CB Singapore. A Random Forest Model accounting for land use mix can predict the activity patterns in off-work recreational hours with up to 50.7% accuracy, but key explanatory variables could still be missing, and other processes could be driving the observed activity patterns. At the time of writing, the Singapore government has announced that the city state will not return to pre-COVID situation in Phase 3, but more activities have resumed continually considering low community cases. Longer periods of data collection on people movement will enable an understanding of how place-visiting behaviours respond to the changing regulations. Bearing in mind the limitations of Twitter data, such as lack of quality control and inconsistent spatial precision, the research questions can be narrowed down further so that it can be adequately and accurately answered with such data.

References

1. Bruns A, Stieglitz S (2012) Quantitative approaches to comparing communication patterns on twitter. J Technol Hum Serv 30(3–4):160–185
2. Davie S (2020) NUS plans to keep students within zones on campus. The straits times, 24 May 2020. https://www.straitstimes.com/singapore/education/nus-plans-to-keep-students-wit hin-zones-on-campus. Accessed 20 Oct 2020
3. Gov.sg (2020) PM Lee: the COVID-19 situation in Singapore (3 Apr). https://www.gov.sg/art icle/pm-lee-hsien-loong-on-the-covid-19-situation-in-singapore-3-apr. Accessed 9 Sept 2020
4. Gov.sg (2020) Ending circuit breaker: phased approach to resuming activities safely. https:// www.gov.sg/article/ending-circuit-breaker-phased-approach-to-resuming-activities-safely. Accessed 31 Oct 2020
5. Gov.sg (2020) Roadmap to phase 3. https://www.gov.sg/article/roadmap-to-phase-3. Accessed 31 Oct 2020
6. Graham M, Hale SA, Gaffney D (2014) Where in the world are you? Geolocation and language identification in Twitter. Prof Geogr 66(4):568–578
7. Hecht B, Hong L, Suh B, Chi EH (2011) Tweets from Justin Bieber's heart: the dynamics of the "Location" field in user profiles. In: CHI '11: proceedings of the SIGCHI conference on human factors in computing systems, pp 237–246
8. Kabir MY, Madria S (2020) CoronaVIS: a real-time COVID-19 Tweets data analyzer and data repository
9. Kovacs-Gyori A, Ristea A, Kolcsar R, Resch B, Crivellari A, Blaschke T (2018) Beyond spatial proximity - classifying parks and their visitors in London based on spatiotemporal and sentiment analysis of Twitter data. Int J Geo-Inf 2018(7):378
10. Lee R, Sumiya K (2010) Measuring geographical regularities of crowd behaviors for Twitter-based geo-social event detection. In: Proceedings of the 2nd ACM SIGSPATIAL international workshop on location based social networks (LBSN '10). Association for computing machinery, New York, NY, USA, 1–10. https://doi-org.libproxy1.nus.edu.sg/
11. Liu X, Long Y (2016) Automated identification and characterisation of parcels with OpenStreetMaps and points of interest. Environ Plann B Plann Des 43(2):341–360
12. Prasetyo PK, Achananuparp P, Lim E (2016) On analysing geotagged tweets for location-based patterns. In: ICDCN' 16: proceedings of the 17th international conference on distributed computing and networking, vol 45, pp 1–6
13. Rao KV, Govardhan A, Rao KVC (2012) Spatiotemporal data mining: issues, tasks and applications. Int J Comput Sci Eng Surv 3(1):39–52
14. Sloan L, Morgan J (2015) Who tweets with their location? Understanding the relationship between demographic characteristics and the use of geoservices and geotagging on Twitter. PLoS ONE 10(11)
15. Soliman A, Soltani K, Yin J, Padmanabhan A, Wang S (2017) Social sensing of urban land use based on analysis of Twitter users' mobility patterns. PLoS ONE 12(7)
16. Statista (n.d.). Number of Twitter users in Singapore in 2019 and 2020 (in millions)*. https:// www.statista.com/statistics/490600/twitter-users-singapore/. Accessed 31 Oct 2020
17. Toh TW (2020) Food delivery sector booms in a time of coronavirus. The straits times. https:// www.straitstimes.com/singapore/transport/sector-booms-in-a-time-of-coronavirus. Accessed 31 Oct 2020

18. Wei Y, Lan M (2015) GIS analysis of depression among Twitter users. Appl Geogr 60(2015):217–223
19. Yan Y, Feng C-C, Huang W, Fan H, Wang Y-C, Zipf A (2020) Volunteered geographic information research in the first decade: a narrative review of selected journal articles in GIScience. Int J Geogr Inf Sci
20. Zhan X, Ukkusuri SV, Zhu F (2014) Inferring urban land use using large-scale social media check-in data. Netw Spat Econ 14:647–667

Space–Time Analytics of New York City Shooting Incidents

Xiang Jing Ang, Hui Ling Wee, Chee Young Goh, and Yingzhe Zhang

Abstract Gun violence in the USA has caused immense socio-economic implications which has plagued urban areas. In the first half of 2020, shooting incidents have been on the rise in New York City (NYC) and a sharp spike of cases was observed in June 2020. In the current study, a space–time analysis on such incidents was done to investigate the clusters and trends with a view to identify NTAs with high shooting densities. Regression analysis on available data for NYC found that such incidents have a close relationship with black population and with locations with vacant housing units. Thus, these factors could play a key role in the predictive analysis of future incidents. A network analysis was conducted which found that shooting incidents generally increase with distance from police stations and that not all of NYC is serviced by police stations within a 15-min drive time. Such findings could potentially help the authorities of NYC to improve their policing, resource allocation and decision-making to address gun violence in the city.

1 Introduction

Gun violence costs America $229 million annually, amounting to $654 per person annually [32]. In New York, gun violence translates to over $2 billion in measurable costs and over $3 billion in pain and suffering annually [18]. Gun violence is a problem that is concentrated in urban areas [22]. Violence tends to cascade into other types of violence making it contagious and diffusive to surrounding neighbourhoods [16, 26]. The number of shooting incidents in New York City (NYC) has been on the rise in the past months; there was a 155% rise in the number of shootings between May 2020 and May 2019, and a 241% rise between June 2020 versus June 2019 [24, 25]. A better insight into the spatiotemporal characteristics and clustering of such

X. J. Ang (✉) · H. L. Wee · C. Y. Goh · Y. Zhang
Faculty of Arts and Social Sciences, National University of Singapore, 1 Arts Link, #03-01 Block AS2, Singapore 117568, Singapore
e-mail: angxiangjing@u.nus.edu

© The Author(s), under exclusive license to Springer Nature Singapore Pte Ltd. 2022
S. N. Kundu (ed.), *Geospatial Data Analytics and Urban Applications*,
Advances in 21st Century Human Settlements,
https://doi.org/10.1007/978-981-16-7649-9_6

shooting incidents could help the authorities to improve their decision making on policing and allocating resources in alleviating the problem.

1.1 Study Area

The current study is limited to the administrative boundary of New York City. The city consists of 5 boroughs—The Bronx, Brooklyn, Manhattan, Queens, and Staten Island and 195 Neighbourhood Tabulation Areas (NTA). The city is highly urbanised city and has a historical problem of gun violence. Moreover, there was a sudden spike in the number of incidents in 2020, with shooting incidents almost doubling as compared to 2019 incidents [4]. Experts posit that the socio-economic and emotional disruption due to the COVID-19 pandemic as one of the key factors in influencing the spike in shooting incidents, but it is impossible to pinpoint a single cause [31].

1.2 Objectives and Scope

The scope of this study is to study the spatiotemporal characteristics of the NYC shootings dataset with an aim to identify trends, hot/cold spots, local outliers and to identify the persistently "problematic" NTAs that has high shooting density. In addition, the relationships between the shootings and selected demographic, socio-economic indicators shall be explored using regression analysis. The service areas of NYPD police stations shall also be investigated in a spatial context to establish if any relation can be drawn on the shooting incidents based on proximity. The study shall discuss the results and provide suitable recommendations to the local authorities for alleviating gun violence.

1.3 Data and Sources

The various datasets used for this study was identified and listed in Table 1. Most of these data were sourced from the government departments of NYC.

2 Methodology

The sourced data were investigated for their attributes, format and their suitability for geographic analysis.

Table 1 Details of data used in the study

S/N	Dataset name	Dataset description	Data source	Data format	Last updated
1	NYPD shooting incident data (Year to Date)	Record of every shooting incident that occurred in NYC from January 2020 to June 2020. The incident's datetime, location and suspect/victim demographics are reported. Data updated quarterly	New York police department (NYPD)	CSV	23/07/20
2	NYPD shooting incident data (Historic)	Record of every shooting incident that occurred in NYC from 2006–2019. The incident's datetime, location and suspect/victim demographics are reported	New York police department (NYPD)	CSV	26/08/20
3	NYC borough boundaries	Boundaries of the boroughs in NYC, excluding any water bodies	NYC Department of City Planning (DCP)	ShapeFile	11/08/20
4	Neighbourhood Tabulation Areas (NTA)	Boundaries of NTA as demarcated using the NYC 2010 census	NYC Department of City Planning (DCP)	ShapeFile	11/08/20
5	NYCHA PSA (Police Service Areas)	Locations of NYPD Police Stations	New York City Housing Authority (NYCHA)	ShapeFile	11/09/18
6	NYC Street Centerline (CSCL)	A representation of NYC's roadbed, with an attribute table containing addresses, road types and traffic directions	NYC Department of Information Technology & Tele-communications (DoITT)	ShapeFile	28/08/20
7	Demographic Profiles of ACS 5 Year Estimates at the NTA level	Selected demographic, economic, housing and social characteristics by NTA from 2012–2016	NYC Department of City Planning (DCP)	XLSX	04/03/20
8	NYPD Arrests Data (Historic)	List of every arrest in NYC going back to 2006 through the end of the previous calendar year	New York Police Department (NYPD)	CSV	16/07/2020
9	Mental Health Service Finder Data	Data used by the NYC Mental Health Service Finder	Office of the Mayor (OTM)	CSV	23/10/2020

2.1 Data Wrangling

Data wrangling is an important step for examining the data. It starts with data discovery and involved the cleaning of data to improve its quality. There were numerous issues identified. Different years of shooting incident datasets are available in multiple files, but the column names of in each file were not standardised. Also, the incident key was found not to be unique, and some incidents in the records were outside of the city jurisdiction area. All these were addressed in the cleaning process. Data which were in textual or tabular format, were transformed into GIS compatible layers for spatial analysis. The whole set of processes are summarised in Fig. 1.

To meet the objectives of the study, two sets of data were prepared. The first one was the unique shooting incident data and the second one was the shooting incident regression layers which contained the explanatory variables. The flowchart for generating these layers are presented in Figs. 2 and 3.

2.1.1 Unique Shooting Incidents

Generally, two datasets were available from the authorities. The historic "NYPD Shooting Incident Data" contained records from 2006 to 2019. The "NYPD Shooting Incident Data (Year to Date)" contained records from January 2017 till June 2020. The data were cleaned to remove redundancy and to obtain a seamless shooting incident dataset from January 2016 to June 2020.

In ArcGIS, the XY Table to Point tool was used to create a point feature class based the location information in the shooting incident dataset. For congruent analysis, the geographic projection used was UTM Zone 18N. Incidents which were located

Data Discovery	**Data Cleaning**	**Data Enrichment**
• Dataset with multiple files	• Standardise column name	• Map the textual records to
• Different column name	• Transform the data to	spatial data, eg. ntacode
• Multiple records with same	correct format, eg.	• Merge dataset with multiple
incident key	occurrence_date	files
• Record falls outside of	• Extract unique incident	• Plot records with spatial
administrative boundary	records	data on to map
• Ambiguous records of	• Remove record that falls	• Aggregate the points layer
perpetrators and victims	outside of study area	to NTA administrative
• Records with null value	• Impute null value with zero	boundary layer
	or NA/unknown	

Fig. 1 Data wrangling methods used in the study

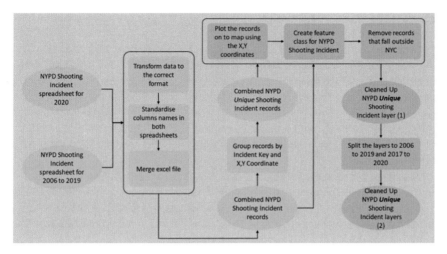

Fig. 2 Flowchart for feature generation methods for NYC shooting incident records

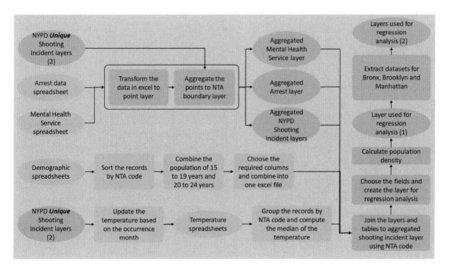

Fig. 3 Flowchart for generating shooting incident regression layers with the explanatory variables

outside NYC were excluded. Two layers were then prepared based on temporal ranges 2006 to 2019 and January 2020 to June 2020.

2.1.2 Shooting Incident Regressions

The arrest and mental health service datasets were transformed into point feature classes which were later aggregated into the NTA boundary polygon layer. The newly generated polygons would now have the attributes for all locations, in addition to the attributes from NTA boundary layer.

From the demographic dataset, the attributes not needed for regression were removed. The data was then sorted by NTA Code (given the NTA Codes are identical in all files). To reduce the data into a single age group category (15 to 24 years), the data for age groups 15 to 19 years and 20 to 24 years were added. Temperature data were updated into the shooting incident dataset based on the occurrence month of the shooting incidents with an aim to compute the median temperature at the time of each shootings. These tables were then joined to the aggregated shooting incident layer based on NTA code.

Similarly, the aggregated Mental Health Service and Arrest layers were also joined to the aggregated shooting incident layer based on NTA code. With the NTA area and population data, population density was calculated for each borough. A subset of the data for 3 selected boroughs—The Bronx, Brooklyn and Manhattan were then prepared.

The above steps were repeated for each shooting incident layer from January 2017 to June 2020.

2.2 Analysis

2.2.1 Exploratory Data Analysis

Using Excel and Tableau, the shooting incident data were analysed to understand its general trends and characteristics. The shooting incidents in NYC exhibited a cyclicity from 2006-June 2020 (Fig. 4). This cyclicity appeared to correlate well with seasons (temperature) as it was observed that shooting incident peak on an annual basis during summer months (June-July) and decrease mostly in the winter months (December-February). This cyclicity was also observed by [21] who associated increased firearm-related violence associated with higher temperatures.

Three time "blocks" exhibited a similar pattern. These blocks were between 2006–2012; 2013–2016; 2017-June 2020 (Fig. 4) and were in a decreasing trend except for the unusual spike May–June 2020 (Fig. 5).

The shooting incidents of May–June 2020 was distinctively higher than that of 2017–2019 and can be construed to be a historical high since 2006. Comparing May–June 2020 to the same period in 2019, there was an increase of 203% in the number of shooting incidents.

From the heatmap (Fig. 6) generated by cross plotting incidents by month and day of the month, it was observed that the highest number of shooting incidents occurred between May–August and are mostly clustered in the second half of the month while the incidents were far less between January and March. This study will focus on two time periods: 2006–2019 and Jan 2017-June 2020. The former time period examines the general trend of shooting incidences, while the latter examines the spike in shootings in May–June 2020.

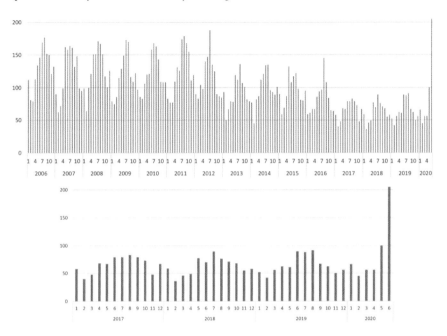

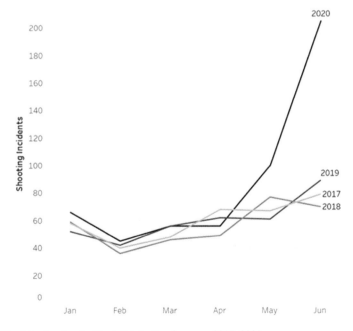

Fig. 4 Monthly shooting incident from 2006—mid 2020 (top) and from 2017—mid 2020 (bottom)

Fig. 5 Monthly shooting incident distribution for years 2017–2020

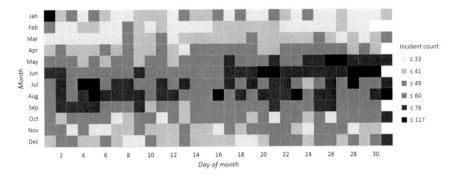

Fig. 6 Incident count cross plot from 2006 to 2020

2.2.2 Shooting Density by NTA

As part of the data wrangling process, the unique shooting incidents were converted into a point feature class and then projected into the chosen geographical projection. The NTA boundary layer was used to summarise the shooting incidents by NTA. Shooting density was then calculated, which was used for the space–time analysis.

2.2.3 Space–Time Analysis

A Space Time Cube (STC), which is a three-dimensional cube containing bins of data points separated by space–time characteristics, was developed. In the STC, the x and y axes represent space, and the z-axis represents time (Fig. 7) [10]. Two STC

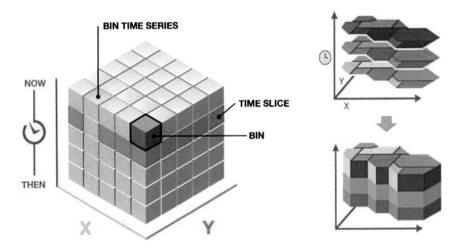

Fig. 7 Space Time Cube illustration [9]

models were created, one for incidents between 2006 and 2019 and the other for Jan 2017-June 2020. The temporal step interval of 1 year was used for the former and 1 month was used for the latter.

The STCs were then visualised for studying the trends based on the Mann–Kendall trend test statistic [14].

Emerging Hotspot Analysis (EHSA) using the STCs was conducted to identify shooting incident hot/coldspots. The results help us visualise the historical clusters with respect to time and one can gain insights as to whether they are intensifying or diminishing clusters, sporadic or oscillating clusters, and consecutive or persistent clusters for each NTA in the time step.

Local Outlier Analysis (LOA) using the STCs was also done to identify statistically significant outliers and clusters with respect to space and time. The result displays the distribution of high-high and low-low clusters, and high-low and low–high outliers.

An 80/20 analysis on the shooting incidents from 2017-June 2020 was done. This analysis was based on the Pareto principle, whereby most of the incidents that occur at a few selected locations, i.e. 80% of the shooting incidents occur at 20% of the locations [9]. The 80/20 analysis identifies clusters from the shootings point dataset and obtain a cumulative percentage field for easy identification of specific locations where there is a disproportionate number of shootings.

2.2.4 Exploratory Regression

The Exploratory Regression tool is a data mining tool which evaluates all the input explanatory variables and selects the best combination of variables for Ordinary Least Squares (OLS) regression analysis [12]. The Exploratory Regression tool accounted for the number of explanatory variables in the OLS model, and the threshold criteria for adjusted R-squared value, coefficient p-values, Variance Inflation Factor (VIF) values, Jarque–Bera (JB) p-values, and spatial autocorrelation p-values on model residuals [12].

The regression layers that were generated from the data wrangling processed was used as an input feature for this exploratory regression. The number of shooting incident by NTA was used as the dependent variables and the selection of explanatory variables are described in the following section. The maximum number of explanatory variables are set to 5 as per the recommendation from ESRI training material [15].

2.2.5 Explanatory Variable Selection

For the explanatory variables, we explored a range of demographic, social, economic indicators by NTA, and eventually narrowed down to the following variables based on academic literature (Table 2).

Table 2 List of explanatory variables

S/N	Variable description	Variable alias	References
1	Black population	BLACK	[6, 28]
2	Number of unemployed individuals	UNEMPLOYED	[27]
3	Number of individuals below poverty	BELOWPOVERTY	[30]
4	Number of vacant house units	VACANTHOUSEUNIT	[19]
5	Rental vacancy rate	RENTALVACANCYRATE	[19]
6	Number of low-education individuals	LOWEDU	[17]
7	Number of foreign-born individuals	FOREIGNBORN	[28]
8	Number of foreign-born and non-citizens	NOTCITIZEN	[28]
9	Number of individuals between the age of 15–24	POP15T24	[3]
10	Median temperature	MEDIANTEMP	[2]
11	Number of mental health service clinics	MENTALHEALTHSVC	[7, 29]
12	Population density	POPDENSITY	[23]

2.2.6 Ordinary Least Squares

The Ordinary Least Squares Tool performs a global OLS linear regression on a combination of explanatory variables to predict or model a dependent variable [13]. In this case, the dependent variable that the study aimed to model was the shooting incidents by NTA. The explanatory variables for the OLS were chosen based on the best passing model from the Exploratory Regression, i.e. model with highest adjusted R-squared and lowest corrected Akaike Information Criteria (AIC). One could utilise the results of the OLS regression analysis to examine correlations and determine the key contributors to the high shooting incidents in the boroughs.

2.2.7 Network Analysis

A Network Dataset was created for NYC based on the NYC road centrelines data. The Service Area was then defined based on reachable times of 5, 10 and 15 min of drive. The number of shooting incidents for each service area drive time class was then calculated to study the variations of incidents for each drive time class.

3 Results and Discussion

3.1 Shooting Incidents in Space and Time

3.1.1 Density and Trends

From the shooting density maps of the two time periods of interest, one can observe that there was a high concentration of shooting incidents in northern Manhattan, northern to south-eastern Bronx, and central to northern Brooklyn (Fig. 8 left). A similar pattern was also observed for January 2017 – June 2020 (Fig. 8 right), which indicated that NTAs with a history of high shooting rates are persistent. Between 2006 and 2019, it was observed that most of the NTAs displayed a downtrend in the time step interval of 1 year (Fig. 9 left), which was commensurate with the findings from the exploratory data analysis. For the period between January 2020 and June 2020, where the change was observed at the finer temporal resolution of 1 month, it was found that only 3 NTAs showed down trends (Fig. 9 right). These were Norwood, Soundview-Bruckner and Soundview-Castle Hill-Clason Point-Harding Park. The remaining 11 NTAs showed uptrends.

Fig. 8 Spatial Distribution of Shooting Density by NTA from 2006–2019 (left) and from 2020–2020 (right)

Fig. 9 Shooting Density Trends by NTA from 2006–2019 (left) and from Jan-June 2020 (right)

3.1.2   Emerging Hotspots

From the Emerging Hotspot Analysis (EHSA) from 2006–2019 it was observed that generally there are more NTAs showing cold spots than hotspots, which was well within our expectation as shooting incidents were on a declining trend. Most of these cold spots were clustered in the NTAs in Queens and Staten Island.

2 persistent hotspots identified were in The Bronx, in the NTAs of Longwood and Melrose South-Mott Haven North (Fig. 10 left). A persistent hotspot means that the location has been a statistically significant hotspot for more than 90% of the time-step interval (1 year), but there was no discernible trend in the intensity of clustering over time [11]. Simply put, these NTAs had consistently higher shooting densities compared to their neighbours and is a subject of further investigation for the local authorities.

There was also a cluster of diminishing hotspots in southwestern Bronx and northern Brooklyn. A diminishing hotspot means that the location has been a significant hotspot for 90% of the time-step interval but the intensity of clustering is decreasing in each time-step [11]. This means that we cannot simply discount a diminishing hotspot even though the shooting densities may be decreasing over time, but the fact that they have higher shooting densities than their neighbours warrant further investigation (Fig. 11).

Between January and June 2020 (Fig. 10 right), albeit a smaller time analysed in a smaller step interval, one can clearly spot the hotspot at the south-central part of NYC. A total of 4 new hotspots could be identified in Brooklyn and 3 in Manhattan. New hotspots are locations that are statistically significant for the final time step interval (June 2020) but has never been a statistically significant hotspot in the earlier time period. The mentioned NTAs experience a sudden rise in shooting incidents which calls for studying the underlying reasons.

Fig. 10 Shooting density hot and cold spots by NTA from 2006–2019 (left) and Jan-June2020 (right)

3.1.3 Outliers

Local Outlier Analysis

From the result of the Local Outlier Analysis (LOA) of 2006–2019 as seen in Fig. 11 (left), it can be observed that most of the NTAs were Low-Low Clusters, commensurate to the decreasing trend of shooting incidents in this time period. A couple of High-Low Outliers, one in Old Astoria and the other in Seagate-Coney Island, could be detected. Again, one can observe that the High-High Clusters are distributed in north-eastern Brooklyn, northern Manhattan and south eastern Bronx, which was in line with the hotspot distribution. The Low–High Outliers were where the parks and cemeteries were located in both Bronx and Manhattan. Locations which show "Multiple Types" meant that the pattern of the shooting density changed over time, and the specific spatiotemporal changes could only be investigated by examining the STC.

For the LOA of January–June 2020 as seen in Fig. 11(right), there was a marked increase in the number of High-Low Outliers. Of these, 6 were in The Bronx, 13 in Brooklyn, 12 in Manhattan, 21 in Queens, and 8 in Staten Island. The south-eastern Bronx had mainly changed into "Multiple Types". North-eastern Brooklyn and part of northern Manhanntan's High-High cluster mostly remains. There were 6 Low–High Outliers—Clinton Hill, Hamilton Heights, Norwood, Prospect Heights, and the parks and cemeteries in Bronx and Manhattan.

On examining the clusters and outliers in the study area, the relevant authorities could investigate the NTAs that was classified as High-Low Outliers or Low-Low Clusters to reason the lower shooting incidents. With possible lessons learnt, policies could be changed and enforced in these NTAs to curb the higher shooting rates in NTAs which were classified as Low–High Outliers or High-High Clusters.

80/20 Analysis

The 80/20 analysis identified shooting clusters where there was a disproportionate number of shooting cases in a few selected locations. Again, a similar pattern where the shooting incidents were clustered in mainly northern Manhattan, northern to south-eastern Bronx, and central to northern Brooklyn was observed. If the local

Fig. 11 Outliers by NTA from 2006–2019 (left) and from Jan–June 2020 (right)

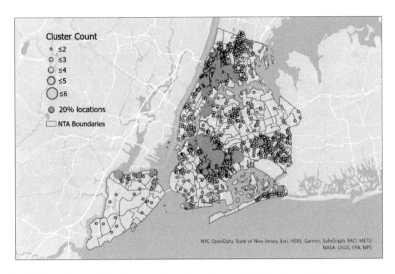

Fig. 12 80/20 analysis of shooting incidents from January 2017 to June 2020

police enforcement could effectively bring down the occurrence of shootings in the highlighted areas (20% locations) in Fig. 12, the city may witnes a significant drop in shooting incidents.

3.2 Regression Analysis

During the initial runs of exploratory regression on the entire NYC, we failed to obtain any passing models. After some rigorous examining from the space–time analysis and regression analysis, exploratory regression on 3 boroughs which experienced high shooting densities/ hotspots was done, instead of a global regression exercise on the entire NYC. The rationales for doing so were.

- a global regression will not be able to capture the nuances in each borough, and we will not be able to compare and investigate the regression on a borough scale;
- it is difficult to capture the entire NYC with a single global regression equation, given that there are different hot/coldspots trends within the boroughs itself.

Multicollinear explanatory variables were removed. E.g. the number of crime occurrence by NTA (from NYPD Arrest dataset) had the same percentage of significance with Black Population.

From the EHSA, it was determined that most of the hotspots were clustered in the boroughs of the Bronx, Brooklyn and Manhattan. Hence, these 3 boroughs were the focus for the regression analysis. OLS regression was used to determine the weights and the causal factors to explain the shooting incident count in these 3 boroughs over the two time periods. The results of the OLS regression are listed in Table 3.

Table 3 Regression equations for the 3 boroughs and 2 time periods of interest

Borough/Time Period	Regression Equation
The Bronx (2006–2019)	COUNT = −20.914140 + 0.004613 * **BLACK** − 0.032627 * UNEMPLOYED + 0.057376 * BELOWPOVERTY + 0.036128 * **VACANTHOUSEUNIT**
The Bronx (January 2017–June 2020)	COUNT = −4.284933 + 0.001089 * **BLACK** − 0.010807 * UNEMPLOYED + 0.006409 * LOWEDU
Brooklyn (2006–2019)	COUNT = 1.678220 + 0.005892 * **BLACK** + 0.047385 * UNEMPLOYED + 0.039576 * BELOWPOVERTY − 0.007882 * FOREIGNBORN
Brooklyn (January 2017–June 2020)	COUNT = 0.328950 + 0.001400 * **BLACK** + 0.009423 * **VACANTHOUSINGUNIT** + 0.003128 * LOWEDU − 0.001742 * FOREIGNBORN
Manhattan (2006–2019)	COUNT = −7.903389 + 0.007293 * **BLACK** + 0.015919 * BELOWPOVERTY − 0.003918 * **VACANTHOUSEUNIT** + 3.054501 * MENTALHEALTHSV
Manhattan (January 2017–June 2020)	COUNT = 17.253758 + 0.001322 * **BLACK** − 0.002397 * **VACANTHOUSINGUNIT** − 2.103953 * RENTALVACANCY + 1.391726 * MENTALHEALTHSV

Table 4 Multiple R-squared values

Borough	Multiple R-squared value for 2006–2019	Multiple R-squared value for January 2017–June 2020
The Bronx	0.903599	0.727384
Brooklyn	0.881173	0.859751
Manhattan	0.959947	0.855993

In the regression equations for the 3 boroughs and 2 time periods of interes, 'COUNT' refers to the count of shooting incidents in that borough for the stated time period.

The multiple R-squared values for the regression equations ranged between 0.73 and 0.95 (Table 4) and displayed a very strong positive relationship between the explanatory variables and shooting incident counts.

From the results of the regression analysis, it was observed that the black population and the number of vacant housing units were important causal factors, which appeared in almost all of the regression results. Each borough was slightly different and required a different set of explanatory variables to explain the observed shooting incidents.

3.3   Network Analysis

The network analysis outputs the service area from NYPD police stations calculated for 5, 10, 15-min drive time (Fig. 13 left). Most of NYC is within a 15-min reach, except for south-eastern Queens, a small area in mid-Manhattan, the north-western tip of The Bronx, and southwestern Staten Island. Shooting incidents from January 2006 to June 2020 that fall within each service area were spatially joined to investigate the way the shooting incident counts changed with distance from service area from the NYPD police stations (Fig. 13 right). It was observed in general that the shooting

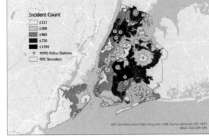

Fig. 13   Time-based service area from NYPD Police Stations (left) and shooting incidents by service areas (right)

incident count was lower in areas proximal to police stations and increased with distance (Fig. 13).

4 Conclusions

Crimes in New York City, including violent crime, have been on the decline since the early 1990s. Despite the historic drop in crime cases, the recent spike in shooting incidents since May 2020 has brought about concerns that the 'bad old days' of violence could be returning. In this study, spatiotemporal data mining techniques in GIS were used to analyse the shooting incidents between for the time periods of 2006–2019 and January – June 2020 to better understand the spatiotemporal characteristics of these events. While it is difficult to pinpoint specific factors that might have led to the sudden increase in shooting incidents, there were sufficient findings from the analysis which could help address the increase.

From the EHSA, two NTAs in Bronx have been identified as persistent hotspots over the long term. This warrants further investigation as the consistent higher shooting densities in these NTAs suggests that there may be systemic factors at play. From the EHSA, four and three new hotspots were identified in Brooklyn and Manhattan respectively over the period of January 2017 to June 2020. These new hotspots coincide with the sudden jump in shooting incidents and should be studied further.

Areas which had lower shooting incidents could provide valuable insights to the law enforcement authority. There were diminishing hotspots in southwestern Bronx and northern Brooklyn which had seen shooting incidents decreasing over the long term. There were also NTAs which had been classified as Low–High Outliers or Low-Low Clusters. The law enforcement authority could look deeper into these areas to elicit possible learning points, such as identifying environmental factors that could be replicated in other "problematic" areas/ NTAs in New York City to help mitigate the rise of shooting incidents.

The regression analysis determined the key demographic and socio-economic factors which influenced the shooting incidents. Amongst the explanatory variables that were correlated to the shooting incidents, the "black population" and "number of vacant housing units" factors were consistent across all the boroughs in New York. As these explanatory variables fall outside the jurisdiction of the police, the city of New York would have to look from a broader perspective and evaluate how the various local government departments could work together to help mitigate the problem of gun violence. The findings from this study has the potential to support authorities of New York City to improve their policies, resource allocation and decision-making processes in their endeavour to alleviate the occurrence of gun violence.

4.1 Limitations

Some limitations of this study would arise out of several factors. The shooting incident locations might not be exact owing to data privacy issues and therefore pose geocoding issues. There were other accuracy issues inherent to the data which could potentially influence the outcomes of this study. Only shooting incidents with injured victims involved were included in the NYPD dataset, otherwise they are classified according to the NYS Penal Law [24, 25]. This would result in our analysis not capturing the full extent of the effect of gun violence in NYC. In addition, there were multiple records of perpetrators and victims captured under a single Incident Key, with no unique identifier for the perpetrator/ victim.

The regression analysis has its own subjectivity too. While it revealed that there were certain factors that were correlated to the shooting incidents, one should be mindful that correlation does not imply causation [1]. Though in 2020 there was a sharp increase in the number of shooting incidents in NYC, it should also be noted that this came amid the backdrop of a historic decline over several years. Experts have not been able to pinpoint a single cause for the rise in shooting incidents or violence, and many have suggested various theories on what is driving up the increase [5, 33]. On the other hand, politicians alleged that the police have purposely slowed down in their work as protest against the changes [33]. In their defence, the police argued that their capability was severely dented due to the budget cuts and a wave of retirements [8]. To the criminologists, there are just too many possible reasons to attribute the increase in crime to. One said the breakdown in trust between the people and police could have played a part while another observed that the loosening of the COVID-19 lockdown measures since June may have created a setting for pent-up crimes after crimes dropped at the beginning of the pandemic [5]. Data for some of these possible factors is either not available, difficult to measure or even not measurable.

On top of this, a recent study from researcher showed that gun purchases led to more gun violence in United States during the pandemic [20]. However, datasets for gun sales and the number of firearm permits issued are not publicly available for the current study.

Despite the limitations, the study is a demonstration of the power spatial analysis and insights that could be drawn t address policing problems for gun related violence in NYC and elsewhere where data is available. The study also highlights the important of collecting and archiving data with integrity and quality good enough to conduct such analyses in the future.

References

1. Altman N, Krzywinski M (2015) Association, correlation and causation. Nat Methods 12(10):899–900. https://doi.org/10.1038/nmeth.3587
2. Anderson CA (1989) Temperature and aggression: ubiquitous effects of heat on occurrence of human violence. Psychol Bull 106(1):74. https://doi.org/10.1037/0033-2909.106.1.74

3. Bushman BJ, Newman K, Calvert SL, Downey G, Dredze M, Gottfredson M, Jablonski NG, Masten AS, Morrill C, Neill DB, Romer D, Webster DW (2016) Youth violence: what we know and what we need to know. Am Psychol 71(1):17–39. https://doi.org/10.1037/a0039687
4. Cable News Network (CNN) (2020) NYC shooting incidents are nearly double this year compared to last year, NYPD stats show. https://edition.cnn.com/2020/09/21/us/new-york-gun-violence/index.html. Accessed 23 October 2020
5. Chapman B (2020) What's fueling New York City's rise in violent crime? There Are Sev Theories Wall Str J. https://www.wsj.com/articles/whats-fueling-new-york-citys-rise-in-violent-crime-there-are-several-theories-11597064288. Accessed 23 Oct 2020
6. Cho JT (2020) Factors affecting crime, fear of crime and satisfaction with police: focusing on policing and the neighborhood context. Polic: Int J 43(5):785–798. https://doi.org/10.1108/PIJPSM-04-2020-0053
7. Dean K, Laursen TM, Pedersen CB, Webb RT, Mortensen PB, Agerbo E (2018) Risk of being subjected to crime, including violent crime, after onset of mental illness: a danish national registry study using police data. JAMA Psychiat 75(7):689–696. https://doi.org/10.1001/jamapsychiatry.2018.0534
8. DeStefano AM (2020) NYPD's head count lowest in 10 years, officials say. Newsday. https://www.newsday.com/long-island/nypd-retirements-shea-1.50031351. Accessed 27 Oct 2020
9. Esri (2020a) 80–20 Analysis (Crime Analysis and Safety. https://pro.arcgis.com/en/pro-app/tool-reference/crime-analysis/eighty-twenty-analysis.htm. Accessed 09 Oct 2020
10. Esri (2020b) Create space time cube from defined locations (Space Time Pattern Mining). https://pro.arcgis.com/en/pro-app/tool-reference/space-time-pattern-mining/createcubefromdefinedlocations.htm Accessed 09 Oct 2020
11. Esri (2020c) How emerging hot spot analysis works. https://pro.arcgis.com/en/pro-app/tool-reference/space-time-pattern-mining/learnmoreemerging.htm
12. Esri (2020d) How exploratory regression works. https://pro.arcgis.com/en/pro-app/tool-reference/spatial-statistics/how-exploratory-regression-works.htm. Accessed 22 Oct 2020
13. Esri (2020e) How OLS regression works. https://pro.arcgis.com/en/pro-app/tool-reference/spatial-statistics/how-ols-regression-works.htm. Accessed 22 October 2020
14. Esri (2020f) Visualizing the Space Time Cube. https://pro.arcgis.com/en/pro-app/tool-reference/space-time-pattern-mining/visualizing-cube-data.htm Accessed 22 Oct 2020
15. Esri (2020g) Introduction to Regression Analysis using ArcGIS Pro. https://www.esri.com/training/catalog/57630430851d31e02a43ee0c/introduction-to-regression-analysis-using-arcgis-pro/. Accessed 22 Oct 2020
16. Fagan J, Wilkinson DL, Davies G (2007) Social contagion of violence. In: Urban seminar series on children's health and safety, John P. Kennedy School of Government, Harvard University, Cambridge, MA, US. Cambridge University Press
17. Fajnzylber P, Lederman D, & Loayza N (2002) What causes violent crime? Eur Econ Rev 46(7). https://doi.org/10.1016/S0014-2921(01)00096-4
18. Gliffords Law Center. (2019) The economic cost of gun violence in New York. https://lawcenter.giffords.org/wp-content/uploads/2018/01/Cost-of-Gun-Violence-in-New-York-1.22.18.pdf. Accessed 04 Sept 2020
19. Groff ER, Lockwood B (2014) Criminogenic facilities and crime across street segments in Philadelphia: uncovering evidence about the spatial extent of facility influence. J Res Crime Delinq 51(3):277–314. https://doi.org/10.1177/0022427813512494
20. Schleimer JP, McCort CD, Pear VA, Shev A, Tomsich E, Asif-Sattar R, Buggs S, Laqueur HS, Wintermute GJ (2020) Firearm purchasing and firearm violence in the first months of the coronovirus pandemic in the United States. https://www.medrxiv.org/content/https://doi.org/10.1101/2020.07.02.20145508v2.article-info
21. Kieltyka J, Kucybala K, Crandall M (2016) Ecologic factors relating to firearm injuries and gun violence in Chicago. J Forensic Leg Med 37:87–90
22. Larsen DA, Lane S, Jennings-Bey T, Haygood-El A, Brundage K, Rubinstein RA (2017) Spatio-temporal patterns of gun violence in Syracuse, New York 2009–2015. PLoS One 12(3)

23. Lee DW, Lee DS (2020) Analysis of influential factors of violent crimes and building a spatial cluster in South Korea. Appl Spat Anal Policy 13(3):759–776. https://doi.org/10.1007/s12061-019-09327-1

24. New York Police Department (NYPD) Shooting Incident Data (Historic). (2020a) https://data.cityofnewyork.us/Public-Safety/NYPD-Shooting-Incident-Data-Historic-/833y-fsy8/data. Accessed 04 Sept 2020

25. New York Police Department (NYPD) Shooting incident data (Year To Date). (2020b) https://data.cityofnewyork.us/Public-Safety/NYPD-Shooting-Incident-Data-Year-To-Date-/5ucz-vwe8/data. Accessed 04 Sept 2020

26. Patel D, Simon MA, Taylor RM (2013) Contagion of violence: workshop summary. National Academies Press, Washington, DC, pp 8–9

27. Raphael S, Winter-Ebmer R (2001) Identifying the effect of unemployment on crime. J Law Econ 44(1):259–283. https://doi.org/10.1086/320275

28. Rosenfeld R, Fornango R, Rengifo AF (2007) The impact of order-maintenance policing on New York City homicide and robbery rates: 1988–2001. Criminology 45(2):355–384. https://doi.org/10.1111/j.1745-9125.2007.00081.x

29. Rueve ME, Welton RS (2008) Violence and mental illness. Psychiatry (Edgmont (Pa. : Township)) 5(5):34–48. https://pubmed.ncbi.nlm.nih.gov/19727251

30. Sozer MA, Merlo AV (2013) The impact of community policing on crime rates: Does the effect of community policing differ in large and small law enforcement agencies? Police Pract Res 14(6):506–521. https://doi.org/10.1080/15614263.2012.661151

31. The New York Times. (2020) N.Y.C. Surpasses 1,000 Shootings Before Labour Day. https://www.nytimes.com/2020/09/03/nyregion/nyc-shootings.html. Accessed 23 Oct 2020

32. United States Joint Economic Committee (JEC) Democratic Staff. (2019) A state-by-state examination of the economic costs of gun violence. https://www.jec.senate.gov/public/_cache/files/b2ee3158-aff4-4563-8c3b-0183ba4a8135/economic-costs-of-gun-violence.pdf. Accessed 04 Sept 2020

33. Zaveri M (2020) A violent august in N.Y.C.: shootings double, and murder is up by 50%. The New York Times. https://www.nytimes.com/2020/09/02/nyregion/nyc-shootings-murders.html

Spatial and Temporal Patterns of Tourist Source Market Emissiveness: A Study of Shanghai, China

Wang Ziwen, Lyu Wenling, Jia Jingnan, and Xi Bohao

Abstract Emissiveness reflects travelers' overall ability to travel from a source to a destination. Shanghai, as a popular tourist city in China, was always a topic for tourism-related analysis. In this study, the emissiveness of 31 provinces in Mainland China as Shanghai's travel source markets were analyzed. At the same time, Travel Gravity Model (Cesario in Econ Geogr 52:363–373, 1976, [1]) was used to incorporate destination attractiveness to construct an advanced emissiveness model. The study found that the source market's propensity to travel to Shanghai showed a decreasing trend from east to west in Mainland China. Besides, a similar popular tourist city, Guangzhou, was chosen to compare the differences in travel propensity between the source markets to the two cities. In terms of time, the travel peaked in the two cities in different months. Spatially, we drew preliminary conclusions that spatial distance and economic development had more important influences on emissiveness.

1 Introduction

Cities are not only important tourist destinations, but also the main source of tourist flow. As tourism enjoys its increasing popularity, the competition of tourist source market has gradually evolved from the competition of tourist destinations because of its significance in tourism marketing and development [2].

1.1 Emissiveness and Tourism

Emissiveness, as a hot research topic of tourism marketing, has been an important criterion for evaluating the underlying tourist source market and travel potential of

W. Ziwen (✉) · L. Wenling · J. Jingnan · X. Bohao
1198 Century Avenue, Pudong, Shanghai 200120, China
e-mail: wang_z_w@icloud.com

© The Author(s), under exclusive license to Springer Nature Singapore Pte Ltd. 2022   121
S. N. Kundu (ed.), *Geospatial Data Analytics and Urban Applications*,
Advances in 21st Century Human Settlements,
https://doi.org/10.1007/978-981-16-7649-9_7

tourist destinations [3]. It is also used in a significant manner to analyze residents' travel intention and potential travel behavior [3].

So far, the research on emissiveness mainly focused on the types, regions, distinctions, source markets and driving factors of emissiveness [4]. According to the differences in research perspectives and scales, emissiveness could be divided into two types: tourist emissiveness and origin emissiveness [5]. The former studies tourists from the micro perspective which are mainly affected by residents' living standard and accessibility opportunities, while the latter studies the sources of recreation visitors from the macro perspective [6].

Research on tourism emissiveness is mostly focuses on the source of tourists. There are few literatures on tourism emissiveness from the perspective of tourist destinations where research data was mostly based on sample surveys and statistical yearbooks. The data used in past research is therefore not very targeted and therefore such studies lacked monthly and quarterly dynamic evolution analysis of travel emissiveness. The current study attempts to overcome that constraint.

1.2 Objective and Scope

The study area for this project was primarily centred around Shanghai but reflects on the whole of Mainland China. This project aims include:

- Estimation of the potential emissiveness of 31 provinces in Mainland China to Shanghai and analyze the temporal and spatial pattern of emissiveness in Shanghai.
- Deduction and usage Travel Demand Index, based on Baidu Index, to analyze differences in time series and temporal evolution of travel demand.
- Spatial analysis of distribution, autocorrelation and hotspot analysis to identify pattern of emissiveness from tourist source market to Shanghai. Besides, tourist score and emissiveness were combined to improve the emissiveness model and derive advanced emissiveness.
- Contrasting the temporal and spatial patterns of the emissiveness model in Shanghai and Guangzhou, in order to justify our conclusions.

1.3 Methodology and Workflow

The focus of the study is on the tourism service of Shanghai vis-à-vis the traveling gravity of 31 provinces in Mainland China with an objective to figure out the salient differences in the spatial and temporal domains. Hence, the structure of the study has three main parts with the first one on Shanghai, the next on the remaining provinces and the final part that reasons the spatio-temporal differences.

The workflow adopted for the study is provided in Fig. 1.

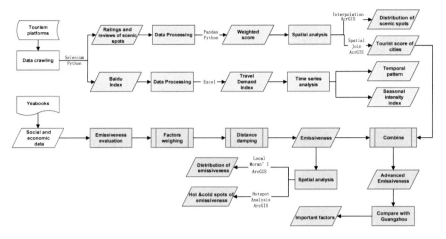

Fig. 1 The workflow adopted for the study

1.4 Definitions

The below definitions are frequently used several times and there it is essential to explicitly define them.

- Tourism Source Market: The tourist source market is the market of potential and existing sources of tourists for a tourist destination [7].
- Travel Demand Index: Using the massive Baidu index to extract the online demand for Shanghai tourism, the travel demand index of Shanghai tourist source markets is abstracted by filtering reasonable keywords.
- Emissiveness: Emissiveness refers to the overall ability of a tourism source market to participate in outdoor recreation or tourism under conditions of travelling ability, travelling willing, distance, transportation condition, etc. [8].

2 Data and Pre-processing

2.1 Yearbook

This study used China Statistical Yearbooks to obtain economic, transportation, and tourism data in various provinces of China, including Shanghai. "China Statistical Yearbook" essential is an annual statistical publication compiled and printed by the National Bureau of Statistics, that comprehensively reflects the economic and social development of the People's Republic of China. The statistical yearbook of a certain year contains a large number of economic and social statistical data of the country and all provinces, autonomous regions, and municipalities in the previous year, as well as major historical years and major national census data of the past two decades.

It is published and issued by the National Bureau of Statistics every year. It is the most comprehensive and authoritative comprehensive statistical yearbook in China. In this study, we select 31 provinces' economic development and traffic condition data to analyze the tourist source destination. Meanwhile, we select data on the level of economic and transportation development of other provinces to evaluate their travel condition.

Four factors were used to build the index of travel condition for each province. These are development level, income level, consumption level, transportation and communication level. We determine the weights of these factors with Entropy Weighted Method (Table 1), and then adjust the weights based on similar research.

Table 1 Travel condition factors and weights

Target Layer	Criterion Layer	Indicators	Weights (%)
Travel Condition	Development Level	Population	3.83
		GDP per person	2.39
		Urbanization rate	1.57
	Income Level	Disposable income per person	4.39
		Disposable income of urban residents per person	1.53
		Disposable income of rural residents per person	5.04
		Average salary of the employees in urban units	3.09
	Consumption Level	Consumption level of urban residents	2.46
		Consumption level of rural residents	5.85
		Total retail sales of consumer goods	5.45
	Transportation and Communication Level	Railway passenger traffic volume	13.85
		Highway passenger traffic volume	13.85
		Air passenger traffic volume	13.39
		Private car ownership	4.64
		Total business volume of postal and telecommunications	5.35

2.2 Scenic Spots

With the development of online platforms, people are more willing to share their travel experiences on the Internet. In China, Ctrip.com is a comprehensive travel website that includes travel services, ticketing services, and travel reviews. And its wide user base ensures the authenticity of its travel reviews. Therefore, our study used web crawler, developed in Python, to obtain the scenic reviews on Ctrip.com.

The main web crawler techniques used was the Selenium package. Selenium was widely used in web crawling according to its working principle of simulating access. By simulating browser access, the JavaScript encrypted content of some web pages was solved. This method had good compatibility and high flexibility but consumed computer memory and was relatively less efficient. Its shortcomings were more obvious in the context of massive data requirements. But while working on small data or subsets of big data, Selenium appears to be more flexible.

In this study, we used Selenium to get the HTML of the web page, and Beautiful-Soup to parse the contents. Then, the reverse geocoding function of the ArcMap API was used to parse the name of the scenic spots and obtain its geographical coordinates. We obtained information on all the scenic spots in Shanghai and Guangzhou. The data was first de-duplicated and invalid uncommented data removed (Table 2).

Secondly, we considered that the higher the number of reviews, the higher was the reliability of the score. With a smaller number of reviews, the scenic spot cannot be considered superior to other attractions even if its score was higher. Due to the large difference in the number of reviews, we used Jenks Natural Breaks Optimization to get breakpoints for classification. This minimized within-group differences and maximized out-group differences. The python Jenkspy package was used to classify the number of reviews in the two regions into three levels by breakpoint: low, average, and high. The scores are stretched according to the different levels to make the distribution of the scores more normally distributed.

$$
weighted\ score = \begin{cases} median(score_{low\ class}) * \log_5 score_{low\ class}, low\ class \\ mean(score_{average\ class}) * \log_5 score_{average\ class}, average\ class \\ score, high class \end{cases}
$$

$$(1)$$

Table 2 Attributes

City	Attributes	Counts
Shanghai	name, href, score (0 to 5), numbers of reviews, longitude, latitude	1370
Guangzhou	name, href, score (0 to 5), numbers of reviews, longitude, latitude	806

2.3   Travel Demand Index

This study measured travel demand from Baidu Index. Baidu Index is a data analysis platform based on Baidu's massive Internet user behavior data, which is considered as one of the most important statistical analysis platforms in China. Every user's search behavior in Baidu provided an indication of active willingness, and the Baidu Index represents online attention, which represented travel demand to a certain extent. Many studies on travel demand have used place names as search terms. This lacks rigor and accuracy as place names don't just mean travel demand, but rather is a kind of network attention. Similarly, we crawled the Baidu index for 31 provinces in China by using Selenium. For this study, we used the keywords of "Shanghai tourist attractions", "Shanghai tourism" and "Shanghai travel tips" to count the search volume of Shanghai users in 31 provinces and stripped out the travel demand index of Shanghai tourist market. The search period was selected to be between January 2016 and December 2018 and results were aggregated on an annual, quarterly and monthly basis.

3   Methods of Analysis

3.1   Spatial Analysis of Destination Attraction

3.1.1   Inverse Distance Weight (IDW)

Using IDW value of an attribute at any point in space could be estimated from discrete data. For an attribute $z = z(x,y)$ at any point(x,y) in space, the inverse distance weighted interpolation formula can be represented as below:

$$\hat{z} = \sum_{i=0}^{n} \frac{1}{d^{\alpha}} z_i \qquad (2)$$

3.2   Travel Demand Index Pattern Evolution

Based on Baidu Index, the evolution of the temporal pattern of Travel Demand index was calculated. The statistics were generated based on temporal intervals as below.

3.2.1 Monthly Travel Demand Index

Keywords like "Shanghai Tourist Attractions", "Shanghai Tourism", "Shanghai Travel Tips" used by used for Baidu search were recovered from the Baidu index to generate the counts for such searches in the temporal range for the study. Searches for Shanghai in 31 provinces across China from 2016 to 2018 were collected to derive the travel demand index of Shanghai tourist market from raw data. Using excel, the data is tallied to finally obtain the tourism demand index for each month from 2016–2018. The three years' data are compared cross-sectionally to find the monthly changes in travel demand index.

3.2.2 Quarterly Travel Demand Index

The monthly travel demand data was summed up to generate the Quarterly Value Index. It was seen that the monthly data distribution pattern of the travel demand index for 2016–2018 is approximately the same. Hence, the quarterly data in 2018 could be analysed.

3.2.3 Seasonality Intensity Index

This index was calculated to reflect the concentration of the time distribution of travel demand in the source markets. The formula for calculating the index is as follows:

$$R = \sqrt{\sum (X_i - \overline{X})^2 / 12} \tag{3}$$

where R represents the seasonal intensity index of travel demands in Shanghai tourist market; X_i is the proportion of travel demand in month i of the whole year; $\overline{X}$ is the average travel demand. The higher the R value tends to zero, the more evenly distributed the index is from month to month; the higher the R value, the greater the seasonal difference in travel demand, and the more obvious the high and low seasons.

3.3 Spatial Analysis of Emissiveness in Tourist Source Markets

3.3.1 Evaluation of Emissiveness

According to the definition and theory of emissiveness, emissiveness includes two core elements: travel demand and the ability to realize travel demand. Besides, the distance can also affect emissiveness. The farther people are located, the less willing they are to travel. Based on these, we can define the emissiveness model as:

$$T_{ij} = A_i D_i exp\left(-\beta r_{ij}\right) \tag{4}$$

where T_{ij} means the emissiveness of the ith province towards Shanghai; A_i means the travel ability of ith province, which is the weighted sum of all the index we collected from yearbooks; D_i means the travel demand of ith province, which is the Baidu Index we have collected; $exp\left(-\beta r_{ij}\right)$ is the space damping, while β is coefficient of space dumping. Here we use $\beta = 0.00322$ [7]. And r_{ij} means the distance between ith province and Shanghai, which can be calculated as the spherical distance of two places:

$$r_{ij} = \frac{\pi R}{180}\arccos(\sin\varphi_i\sin\varphi_j + \cos\varphi_i\cos\varphi_j\cos\left(\lambda_j - \lambda_i\right)) \tag{5}$$

where $\pi = 3.14$; R is the earth radius and $R = 6371 km$; (φ_i, λ_i) and (φ_j, λ_j) represent the geographical coordinates of place i and place j respectively (φ is latitude, while λ is longitude).

3.3.2 Advanced Emissiveness Model

The emissiveness towards tourist destination was also related to the tourist attractions the destination owns. Different places might have different tourist attractions. Therefore, the emissiveness model was improved by adding the scores of the tourist destination, so that different tourist destinations emissiveness of the tourist source market could be compared.

The advanced emissiveness mode was:

$$AT_{ij} = A_i D_i exp\left(-\beta r_{ij}\right)S_i \tag{6}$$

where AT_{ij} is the advanced emissiveness, and S_i is the review score for each tourist destination.

3.3.3 Spatial Autocorrelation

Spatial autocorrelation indicates the degree of connection between things at different distances. According to different measurement scales, it could be divided into two types: global and local. The commonly used metric for spatial autocorrelation is Moran's I, while the local spatial autocorrelation can be divided into high and low clusters by G_i^* index:

$$I = \frac{\sum_{i=1}^{n} \sum_{j=1}^{n} w_{ij}(x_i - \overline{x})(x_j - \overline{x})}{\left(\sum_{i=1}^{n} \sum_{j=1}^{n} w_{ij}\right) \sum_{i=1}^{n}(x_i - \overline{x})^2} \tag{7}$$

$$G_i^* = \frac{\sum_{j=1}^{n} w_{ij}x_j - \overline{x}\sum_{j=1}^{n} w_{ij}}{\sqrt{\frac{\sum_{j=1}^{n} x_j^2}{n} - (\overline{x})^2}\sqrt{\frac{n\sum_{j=1}^{n} w_{ij}^2 - \left(\sum_{j=1}^{n} w_{ij}\right)^2}{n-1}}} \tag{8}$$

where $\overline{x}$ is the average of emissiveness in this study; n is the number of provinces; w_{ij} is the spatial weight matrix. Moran's I range from -1 to 1. If I > 0, then the emissiveness shows positive spatial autocorrelation with the clustered places, which means that the more clustering places will have greater emissiveness; if I < 0, then the emissiveness shows negative spatial autocorrelation with the clustered places.

4 Results

In the first part of the study, the spatial distribution of scenic spots in Shanghai and Guangzhou were analysed and their average scores in terms of districts as the tourist attractiveness of the city was discussed.

4.1 Spatial Distribution and Tourist Score

In this section, the spatial distribution of the attraction scores with IDW spatial interpolation was performed. Scenic spots are discrete features, but scenic spots can influence the surrounding tourism supporting facilities' service ability which can be continuous features (Fig. 2).

Shanghai Disney Land had driven the demand for accommodation in the surrounding areas. Many homestays had emerged in Shanghai Disneyland and surrounding rural areas. At the same time, these homestays tend to upgrade their accommodation conditions and create their own unique homestay features to attract more tourists. Hence, these homestays have become not only for convenient accommodation near Disneyland, but also have generated new tourist destinations for weekend recreation. Therefore, the service ability cannot be just measured by a scenic spot score. In summary, the interpolation used here was to find out the distribution pattern of scenic spots' score for having a description of the tourism service ability. The interpolation result was not used to predict the scenic spots' score in other areas without a given score.

From Fig. 2, Guangzhou and Shanghai's have the highest value tourist ratings and points are spread across all districts. Overall, Guangzhou's scenic spots are rated higher than Shanghai's. In the central part of Guangzhou, Yuexiu, Tianhe, Baiyun, and

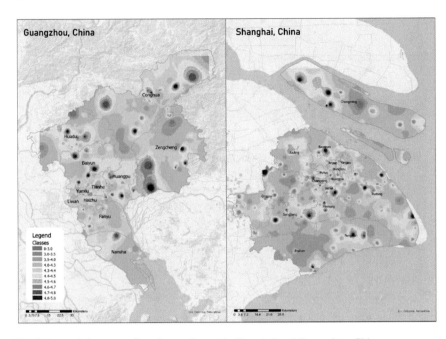

Fig. 2 Interpolation map of scenic spots' score in Shanghai and Guangzhou, China

Huadu, clustering of high score values are observed. In Shanghai, in addition to high value clustering in the central northeast corner, there were localized significant high values in the Pudong and Songjiang districts, which are related to the large scenic spots Shanghai Disneyland and Shanghai Happy Valley. In contrast, Guangzhou Panyu District, which has the similar type of attraction, Changlong Tourist Resort, did not show significant high values.

At the district scale, the scenic spots of the city were connected and summarizing. An average score for each district and for the city as a whole was calculated by the following formula:

$$\text{Tourist score} = \frac{\sum_{spots_{d_i}} weighted\ score * numbers\ of\ reviews}{\sum_{spots_{d_i}} numbers\ of\ reviews} \tag{9}$$

The density distribution of scenic spots throughout the city was generated and overlaid on the average tourism score of the district (Fig. 3).

From the clustering of density of scenic spots (Fig. 3), it was concluded that Yuexiu district in Guangzhou province and Huangpu and Hongkou districts of Shanghai had the most scenic spots. In terms of district tourism scores, Guangzhou Panyu District had the highest score, which was primarily because of the Changlong Tourist Resort, which is the district's large tourist attraction in the district. Yuexiu District, which despite having a concentration of scenic spots scored lower. Shanghai's high

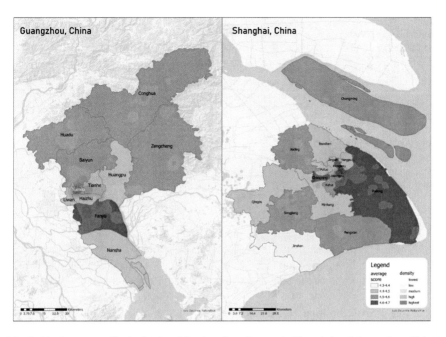

Fig. 3 Tourist score at district level and density of scenic spots in Shanghai and Guangzhou, China

tourism score was concentrated in Hongkou District, Changning District, and Pudong District, which are high value areas in terms of attraction density. At the city level, Shanghai has an average review score of 4.576, while Guangzhou's average review score was 4.596.

4.2 Travel Demand Index Patterns

Based on the analysis of the monthly travel demand data for the three years 2016–2018 (Fig. 4), it was found that the peak of the travel demand index was between July

Fig. 4 Monthly changes in travel demand index

Table 3 Quarterly travel demand index of Shanghai tourist market, 2018

Quarter	Travel demands
First quarter (Jan-Mar)	241,935
Second quarter (Apr-Jun)	215,731
Third quarter (Jul-Sept)	329,598
Fourth quarter (Oct-Dec)	181,692

Table 4 Seasonality intensity index of Tourism demand in Shanghai, 2016–2018

Year	R(%)
2016	1.96
2017	2.22
2018	2.96

and August, while there was a small peak of travel demand in January–February. At the same time, comparing the three-year data (Fig. 4) revealed that the relationship between the volume of the annual travel demand index was incremental from 2016 to 2018.

Table 3 lists the quarterly travel demand figures. The first and third quarters displayed higher demand for travelling in Shanghai's tourism market (with the third quarter being the highest).

By calculating and comparing the R-values for each year from 2016–2018 (Table 4), it was observed that the R-values are increasing year on year, which meant that the seasonal differences in tourism demand in Shanghai's tourism market were more and more obvious. From 2016 to 2018, the seasonal differences in travel demand in the Shanghai source market grew year on year. Over time, people were more willing to choose the months based on suitable weather and travelled more to Shanghai. The travel demand peaked in July.

Travel Demand Index comparison between different cities were done and results presented in Fig. 5 and Table 5. Comparing the temporal changes in tourism demand in the two cities, one can see that the monthly changes in travel demand index in Guangzhou was also seasonal, but the seasonal effect was not as pronounced as it was

Fig. 5 Monthly changes in travel demand index in Shanghai and Guangzhou

Table 5 Quarterly travel demand index in Shanghai and Guangzhou

Quarter	Shanghai travel demands	Guangzhou travel demands
First Quarter (Jan-Mar)	241,935	194,737
Second Quarter (Apr-Jun)	215,731	170,580
Third Quarter (Jul-Sept)	329,598	205,206
Fourth Quarter (Oct-Dec)	181,692	155,816

for Shanghai. Besides, the quarterly travel demand index showed that third quarter was the peak of tourism both in Shanghai and Guangzhou.

4.3 Emissiveness in Tourist Source Markets

Figure 6 shows the distribution of emissiveness for 31 provinces in Mainland China. Coastal provinces to the east have greater emissiveness than inland provinces in the western parts. In particular, Jiangsu, Zhejiang and Shandong provinces have higher emissiveness. Overall, there were clear spatial differences in emissiveness which had a strong correlation with population density. The divide line between east and west matches with "Heihe-Tengchong Line" (the well-known contrast line of population density in China). This essentially indicated that the eastern of country where most of the Chinese population reside, had a greater emissiveness than the western parts that had less population (Table 6).

Cluster analysis of emissiveness (Fig. 7) shows that both Jiangsu and Zhejiang, which is located adjacent to Shanghai, and surroundings had high emissiveness. However, Anhui, despite being closer to Shanghai, appeared to be cold spot cluster with low–high emissiveness. On the other hand, Guangdong presents a high-low outlier. The surrounding provinces of Guangdong had lower emissiveness towards Shanghai, but in contrast, Guangdong was the one with highest emissiveness.

According to the spatial distribution of emissiveness and the evolution of cold-hot spots, Shanghai's tourist source market could be divided into four categories: strongest emissiveness area, stronger emissiveness area, general emissiveness area, and weak emissiveness area (Fig. 8). The map does not include Taiwan, Hong Kong and Macau for lack of data and the South China Sea islands were included in Hainan and Guangdong provinces for simplified cartography.

The strongest emissiveness area covered the four provinces of Shanghai, Jiangsu, Zhejiang and Shandong. These provinces were all located in the east coast, where the income and consumption levels of residents were relatively high. In addition, these provinces were adjacent to Shanghai, that strengthens the emissiveness in terms of proximity.

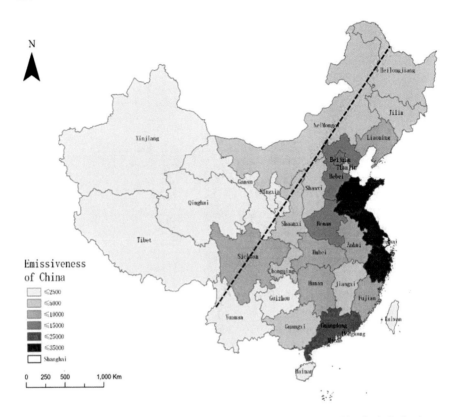

Fig. 6 Emissiveness by provinces in Mainland China. Heihe-Tengchong Line in dashed red

The stronger emissiveness area included five provinces of Guangdong, Henan, Beijing, Fujian, and Hebei, which could be categorized into two. The first was the dominant economic development which includes Beijing and Guangdong and had no advantage in terms of travel cost. But as the level of economic development was generally high, the ability to travel was strong too. The second was where geographical proximity dominated. These areas are represented by Henan and Fujian, which are relatively close to Shanghai making travel relatively convenient. Besides, the economic development of these places was not so lower than the first.

The general emissiveness covered 9 provinces and was mainly distributed around Central and North China. Economic development and traffic conditions at these regions, represented by Shaanxi and Tianjin, are medium.

The weak emissiveness area included 13 provinces, located mostly in the north-west and south-west regions where economic development level is relatively low. The representative provinces are Qinghai and Tibet. Besides, these regions are far from Shanghai. High travel costs and low travel convenience restricts the ability for people from these regions to Shanghai.

Table 6 The proportion of tourist market emissiveness in Mainland China

Province	Proportion (%)	Province	Proportion (%)
Shanghai	0.2881	Tianjing	0.0132
Jiangsu	0.0914	Jiangxi	0.0131
Zhejiang	0.0897	Heilongjiang	0.0118
Shandong	0.0819	Jilin	0.0109
Guangdong	0.0698	Chongqing	0.0103
Henan	0.042	Shanxi	0.0099
Beijing	0.037	Guangxi	0.0096
Hebei	0.0361	Yunnan	0.0079
Fujian	0.0229	Xinjiang	0.0072
Liaoning	0.0209	Gansu	0.0065
Sichuan	0.0188	Guizhou	0.0065
Anhui	0.0176	Hainan	0.0048
Hubei	0.0174	Ningxia	0.0044
Hunan	0.0166	Qinghai	0.0017
Shaanxi	0.0162	Tibet	0.0013
Inner Mongolia	0.0147		

Fig. 7 Clustering of Emissiveness in China

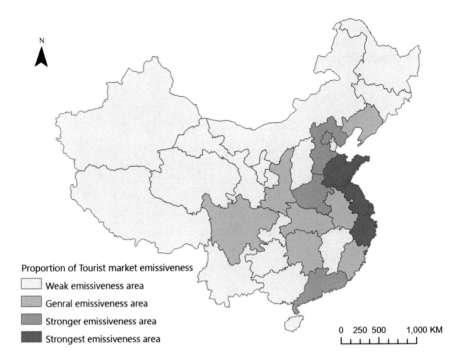

Fig. 8 Proportion of Tourist market emissiveness for 31 Provinces in Mainland China

Comparing Guangzhou against Shanghai, the review scores being at 4.596 and 4,576 respectively, are relatively similar. But the tourist source markets for these locations appear to be very different. Distance seems to be an important factor which impacts people's emissiveness towards these two cities. Shanghai is a more attractive for its surrounding provinces, such as Jiangsu and Zhejiang. The population of central parts of the country, which is closer to Guangzhou, are more likely to travel to Guangzhou than to Shanghai, while people in the northern of country prefer to travel in Shanghai because of proximity. Besides, the economic level of each province too appeared to have a greater influence on the emissiveness. Economically developed provinces like Beijing, Guangdong and Zhejiang seem to be rarely affected by distance as people can choose to travel farther destinations.

5 Conclusions

The conclusions of this study can be summarised as below:

- Tourist destination attractiveness is not just related to the number of scenic spots.
- Temporal pattern of tourist demand was both general and specific.

- Emissiveness from tourist source market to Shanghai was found to have a regional convergence.
- Space distance and factors of economic level are the prime factors of emissiveness of tourist source market.

A city with more scenic spots does not necessarily mean higher tourist destination attractiveness. Shanghai has more scenic spots, but its tourist destination attractiveness is slightly lower than Guangzhou. Besides, city centre in Shanghai and Guangzhou tends to have higher scenic spot density but do not garner a high score.

Among Tourist Demand Index of 2016–2018, the summer vacation (generally July–August) and the Chinese New Year (generally January–February) in mainland China were the peak periods of travel demand. During Summer vacations, students and parents were willing to travel, while during Chinese New Year, family travel becomes the main reason. Travel demand peaks of Shanghai usually are in the month of July. Travel Demand Index were found to be increasing from 2016 to 2018 but they displayed seasonality across the years because of vacation months.

On the regionality of emissiveness, it was not difficult to conclude that the eastern coastal and economically developed cities had higher emissiveness than the west. Cities in Central China have general emissiveness, while the inland southwest and northwest provinces have weak emissiveness primarily due to the constraints of transportation costs and economic development.

Comparing Shanghai and Guangzhou with the advanced emissiveness model, distance and economic level was found to be the main factors influencing emissiveness. Even though the two cities have similar attractiveness, people preferred to travel to nearby places instead of travelling far away. The provinces which have better economic development always had higher emissiveness.

The emissiveness towards tourist destination is also related to the tourist attractions the destination it hosts. Therefore, it is rather one-sided to only use tourism demand to describe the subjective wishes of tourists. Hence the Travel Gravity Model was used to improve the traditional emissiveness model. By analyzing the scenic spots distribution and tourists' reviews, one can generally derive the attractiveness of the tourist destination with a reliable score. The score index of the tourist destination was helpful in improving the emissiveness model, which can be used to compare different tourist destination's emissiveness of tourist source market.

5.1   Limitations

The use of Baidu Index to derive the Travel Demand Index has several limitations as it is based on keyword searches. It is not possible to include all keyword to extract the score. Moreover, only a fraction of travel willingness is converted into actual travel. Another limitation is the use of review numbers and score from Ctrip.com. Other factors such as traffic condition and consumption level which are likely to influence tourist destination attractiveness was not included in this study.

5.2 Future Scope

One can improve the accuracy of advanced emissiveness model by adding more related factors, thereby making the advanced emissiveness model more comprehensive. It is also possible to normalize the tourist destination attractiveness by deriving and including more cities' tourist destination attractiveness. Therefore, a potential future scope would be to include more causative factors and include more destinations for a more inclusive emissive model.

References

1. Cesario FJ (1976) Alternative models for spatial choice. Econ Geogr 52(4):363–373
2. Shi C, Zhang J, You H (2006) Spatial disparities of latent emissiveness of urban residents in China. Sci Geogr Sin 26(6):628
3. Lai S, Zou Y (2016) The evaluation of tourism area's origin emissiveness: a case study of Gutian tourism area in Fujian. Econ Geogr 36(3):194–200
4. Chen C, Xie H (2006) Comprehensive evaluation of the regional diversity of latent domestic emissiveness (lDE) in Fujian province. Econ Geogr 26(5):884–887
5. Zhong SE, Zhang J, Ren LX, Luo H, Li M, Dong XW (2009) A study of the provincial emissiveness in China based on the socio-economic properties– with contrast between China and other countries. Econ Geogr 29(1):153–158
6. Leiper N (1979) The framework of tourism: towards a definition of tourism, tourist, and the tourist industry. Ann Tour Res 6(4):390–407
7. Li S, Wang Z, Zhong ZQ (2012) Gravity model for tourism spatial interaction: basic form, parameter estimation, and applications. Acta Geogr Sin 04:536–544
8. Wu BH (1997) Emissiveness and destination choice behavior of Shanghainese in their weekend recreation. Hum Geogr 12(1):17–23

Spatial Perspectives of Crime Patterns in Chicago Amid Covid-19

Shuhan Yang, Soomin Kang, Sharon Low, and Lei Wang

Abstract Data from Chicago city in Illinois State of the U.S. was extracted to estimate the effects of the onset of the COVID-19 pandemic on crime. There was a general drop in reported crimes, which appeared to precede the stay-at-home orders, and then there was a sudden increase in non-residential car theft after the stay-at-home order issued in March. This change suggested that criminal activity was "substituted" as most people would be staying at home with minimal surveillance of the vehicles which were parked elsewhere. On the contrary, there was an immediate increase in domestic crime from January onwards. This study aims to figure out the impact of the pandemic on changing trends in crime by running various GIS methods. As a result, monthly change of spatial pattern, change in domestic crime and motor theft, and the relationship between some socioeconomic factors and crimes were analyzed. Evidence showed that there was a need to establish an association between crime rates with social and health data to ensure adequate investment in social safety net and programs to strengthen social resilience given that the pandemic was likely to continue for months if not years.

1 Introduction

The outbreak of novel coronavirus pandemic in the People's Republic of China (PRC) since December 2019 and in the United States (U.S.) since January 2020 had profoundly impacted on society and economy [1]. Institutions were responding to the pandemic through one of the strictest lockdowns in global history, which severely affected economic activity, and the consequent reshaping of socio-economic norms affects how criminals operate [2]. Additionally, women and children, as well as physically and mentally challenged are among those who are most vulnerable and unsafe during this pandemic. This is because they might have to stay with people who could cause hurt during the imposed lockdown [3].

S. Yang (✉) · S. Kang · S. Low · L. Wang
National University of Singapore, 1 Arts Link Kent Ridge, Singapore 117570, Singapore
e-mail: yangshuhanaz@163.com

© The Author(s), under exclusive license to Springer Nature Singapore Pte Ltd. 2022   139
S. N. Kundu (ed.), *Geospatial Data Analytics and Urban Applications*,
Advances in 21st Century Human Settlements,
https://doi.org/10.1007/978-981-16-7649-9_8

Fighting and adapting to the pandemic could be costly. While these effects are "unprecedented" and might be difficult to fully predict, what was apparent was that some criminal activities were reduced due to lack of opportunity and at the same time other crimes have emerged and increased. The realignment of resources—ranging from healthcare, education to public services, including law enforcement services in responding to the pandemic, may have significant influences on how such services and groups evolve in the months to come [4].

Globally, the literature on the effects of COVID-19 pandemic on crime is scarce, given the brief time that has passed since it began. Several papers reported mixed results from minor changes to substantial declines across major cities in the United States [5], such as New York, San Francisco, Los Angeles [6], Chicago, and Philadelphia to Queensland, Australia [7] and Lancashire, United Kingdom [8]. Crime is hard to explain even in normal times. Some of the trends this year were relatively straightforward [9]. Residential robberies declined with people spending more time at home. Shoplifting declined when businesses are closed. Stolen cars increased in some cities as novice delivery drivers started leaving their running cars on the road. Two recent research which focused specifically on the effect on domestic violence,[10] reported an increase of 7.5% in the first several weeks after the onset of the pandemic but [11] found weak evidence for such an increase.

In this study, some underlying assumptions were drawn for how reported crime may be affected by the onset of the pandemic. Reported crime incidents depend not just on the true crime rate, but also on the reporting rate [12]. For most crime types, other than drugs, public or victim, reports are the main source of detection [12]. Given the movement restrictions imposed as a preventive measure of COVID-19, the probability of observation by the public drops. However, there is also some evidence that police presence and enforcement of certain laws decreased during the pandemic, which would tend to increase crime rates due to the decreased expected penalty [13]. Thus, one should expect to see an increase in property crimes like car theft, theft from vehicles, and non-residential burglaries because of the decline in expected penalty due to a drop in mobility [5]. On the other hand, with much more time spent at home, one might expect an increase in domestic violence.

1.1 Study Area

The City of Chicago is the most populous city in the State of Illinois and the third-most-populous city in the United States. As of 22 October, there have been 91,589 cases of COVID-19 identified in Chicago residents and 360,159 in Illinois. Concerns of a crime wave have been raised primarily by law enforcement agencies. The pandemic's rapid spread has required them to quickly respond to the crisis. According to the Center for Criminal Justice Research, Policy, and Practice of the Loyola University Chicago [14], the Covid-19's containment effect on Chicago crime trends reported a statistically significant decrease in narcotics, criminal trespasses, deceptive practice, theft, battery whereas weapons violation was statistically significantly increased. Other reported crimes like assault, criminal damage, robbery, burglary, and motor vehicle theft were largely unaffected by COVID 19 stay-at-home orders.

Due to the availability of data, this research focused the analysis period as within January 20th, 2020 when the first infected patient was reported in the United States to May 25th, 2020 when overall reported trends in crimes changed dramatically after George Floyd was killed [14]. The trend comparison was done with the two period, before and after COVID-19, which was from September 17th, 2019 to January 19th, 2020 and from January 20th, 2020 to May 24th, 2020.

2 Methodology

Spatial analysis for crime mapping was used as a potential tool to guide effective law enforcement and crime management. Based on the availability of open-source data, Chicago city in Illinoi State was used as a case study.

Heat maps were generated using kernel density estimation (KDE), was helpful in estimating the unknown cell values considering the neighbouring observed values. In addition, hot spot analysis and density-based clustering were used for identifying clusters. Hot spot analysis is the best-known way of pinpointing the location and the extent of clusters with spatial statistics. Meanwhile, density-based clustering is a functional method for discovering clusters, based on the density-based algorithm proposed by [15].

For the last part of the analysis, an Ordinary Least Square (OLS) regression and Moran's I were performed for figuring out the relationship between crime cases and demographic factors and for validation of the model. To achieve this, crime rates instead of frequency were used as the dependent variable as it can better reflect the level of crime risk faced by a specific area of group of population. The crime rate was estimated per 10,000 people. The calculation formula is as follows:

$$Ri = 10000 * Ci/Pi$$

where Ri is the crime rate of community i. Since Chicago contains 77 communities, the value range of i is from 1 to 77. Ci is the total amount of crime in community i within a period; Pi is the total number of residents in the community i.

Due to the paucity of social-economic data at the community level, this study selected five (5) explanatory variables as proxy indicators to reflect the population, economy, employment, and residence using secondary data to conduct the regression analysis with crime rates. The summary of the explanatory variables is reflected in Table 1.

OLS regression was the global regression model which provided insights for the factors behind observed spatial patterns by creating a single equation for the whole study area. To ensure that the residuals derived from the regression model

Table 1 Explanatory variables used for regression analysis

Variables	Unit	Explanation
Population density	People/km^2	Measure the density of a community's population
Below Poverty Level	%	Measure the level of poverty in a community
Per Capita Income	thousand US dollars	Measure the income level of the residents in a community
Unemployment	%	Unemployment rate
Crowded Housing	%	The ratio of occupied housing units of more than one person per room to all housing units

were randomly distributed over space, Moran's Index was calculated before the interpretation of the result.

The analysis showed areas of crime hotspots, areas requiring constant police patrol, and new crime types which citizens are likely to faced post-pandemic period. The process of the analysis is shown in Fig. 1.

In summary, the project's overall workflow can be divided into these three major parts:

Part 1: Creating heat maps by KDE to grasp the monthly change in spatial patterns of crime.

Part 2: Mapping clusters of crime types before and after pandemic by running hot spot analysis and density-based clustering, focusing on motor theft and domestic violence. For both periods before and after the pandemic, a hot spot analysis and a density-based clustering were processed in parallel to determine hot spot areas and better understand the result of the analysis.

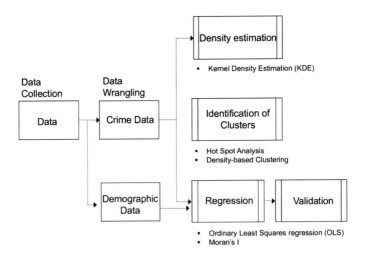

Fig. 1 Workflow for the analysis

Part 3: Running OLS to find out which of the socio-economic factors of the population had a significant relationship with the crime to predict future crime trend for the post-pandemic period. Population density, Below Poverty Level (%), Per Capita Income, Unemployment (%), and Crowded Housing (%) were used as explanatory variables.

All analysis was conducted using ArcGIS Pro. Incidence data were collected from Chicago Data Portal (https://data.cityofchicago.org/) and socioeconomic data were collected from Data USA (https://datausa.io/).

3   Analysis and Results

3.1   Crime Trends

This project tracks the effect of COVID-19 containment policies on crime trends in Chicago monthly. Line chart and KDE were used to present the temporal and spatial trends.

Figure 2 revealed the trend of reported total crimes. The steep decline between March and April is likely the effects of the statewide Executive Order for Illinois residents, including Chicago, to stay at home for a period of three (3) weeks from March 20th to April 7th [16].

The spatial autocorrelation analysis was carried out on the crime point data of the two time periods before and after the onset of COVID-19. Table 2 shows that Global Moran's I of Period A was 0.035, and the Z score was 19.307. The probability of random generation of this clustering pattern was less than 1%, and the probability of data clustering was greater than the probability of random distribution.

On the other hand, the Global Moran's I of period B was 0.007, and the Z score was 2.498. The probability of random generation of this clustering pattern was less than 5%, and the probability of data clustering was greater than the probability of random distribution.

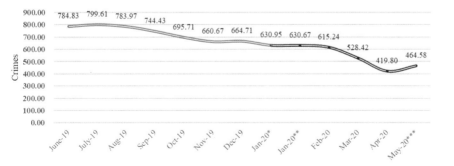

Fig. 2   Crime Trend from June 2019–May 2020. Pre-Covid (green) and during Covid (red)

Table 2 Autocorrelation report

	Period A (Before COVID)	Period B (During COVID)
Global Moran's I	0.035	0.007
Variance:	0.000	0.000
z-score:	19.307	2.498
p-value:	0.000	0.012

These results indicate that the distributions of crime cases before and after the COVID-19 in Chicago have significant spatial clustering and a positive spatial correlation pattern.

KDE (Fig. 3) revealed a more interesting pattern of crimes in Chicago. Prior to the onset of COVID-19, the spatial distribution of crime in Chicago was clustered mainly in downtown Chicago, such as Loop, Near North, Near South, with smaller clusters in the western part of the city like Austin, Garfield Park, North Lawndale. Whereas after onset of pandemic, the clustering of crimes in urban centers declined, and the western part of the city gradually became more of a crime hotspot over the months. In addition, several areas in the southern part of the city like Greater Grand Crossing, Woodlawn, South Shore was observed to be an area of concern specifically from April onwards.

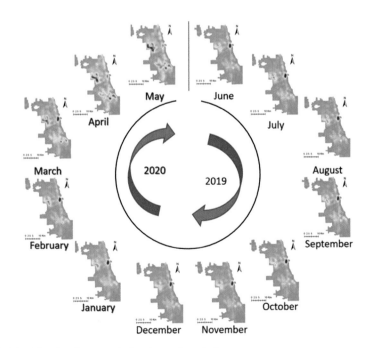

Fig. 3 Monthly crime hotspots during the study period

3.2 Trends by Type

3.2.1 Motor Thefts

Figure 4 revealed a sudden drop in motor theft in April and then a steep increase in May. This observation was similar to the overall trend of reported crimes in Fig. 2. This increase could be due to the anticipation of reduced police presence and enforcement of certain laws during the movement restrictions imposed by the Illinois Governor.

According to Fig. 5, the dark blue cluster of reported motor theft crimes around the Albany Park shifted to Lake View after the outbreak. The green cluster expanded

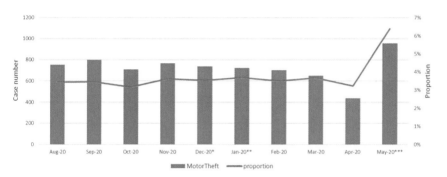

Fig. 4 Monthly motor thefts and Proportion with respect to total crimes

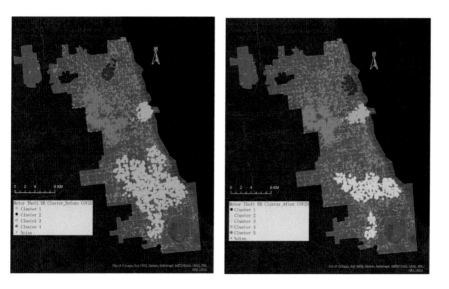

Fig. 5 Density clusters of motor theft events before Covid (left) and after Covid (right)

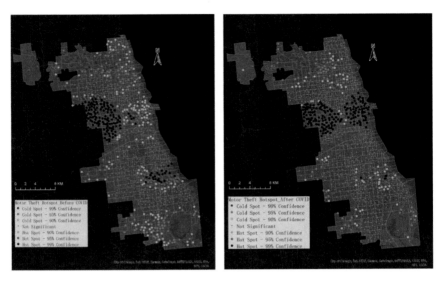

Fig. 6 Motor theft hotspots before Covid (Left) and after Covid (right)

from Loop and Near North Side areas towards Near West Side whereas the yellow cluster before COVID-19, centred around Garfield Park, was shortened. And the biggest light blue cluster in southern Chicago broke into two after COVID-19.

Hotspot analysis of the reported motor theft crimes (Fig. 6) were observed in the similar cluster areas as DBC (Fig. 5) and revealed more details.

Before COVID-19 onset, hotspot clusters were observed around Loop and Near North Side (right side of map), Garfield Park (left side of the map) and smaller cluster in the southern Chicago. The onset of COVID-19 saw the hotspot increased in Loop and Near North Side and reduced in southern Chicago. A more in-depth review of the community characteristics of the neighborhood revealed that COVID-19 had a negative impact on the global economy, including Chicago families and companies [17], some of whom living on the edge might be tempted or lured to commit criminal offences when it was perceived that police and public presence were declined and chances of being caught were reduced. As for communities in southern Chicago, on the Chicago Neighborhood Map (https://hoodmaps.com/chicago-neighb orhood-map) where *'never live here'* were tagged around Garfield Park and *'hell on earth,'* *'murders'* and *'dangerous'* were tagged around Austin and Humboldt Park, indicating the complicated crime environment. Where life-threatening crime was of high possibility in occurring, COVID-19 may lead to more violence such as weapon violation instead of theft.

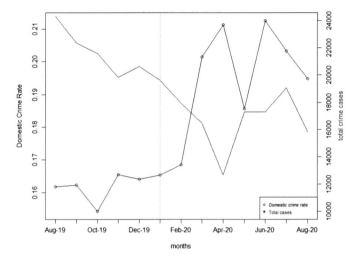

Fig. 7 Domestic crimes against total crimes during the study period

3.2.2 Domestic Crimes

This project further tracked the effect of COVID-19 on domestic crime trends on a monthly basis. For this purpose of the analysis, crime data were re-coded as 'domestic' and 'non-domestic' crime. Figure 7 showed that the trend of domestic crimes surged from February onwards, even though the overall trend of reported crimes went down since the onset of COVID-19.

Distributions of domestic and non-domestic crime hotspots were established for each period using hotspot analysis. The spatial patterns in Fig. 8 showed a clear difference between domestic crime and non-domestic crime. Prior to the on-set of COVID-19, the hot spot of non-domestic crimes was strongly concentrated in downtown Chicago, whereas the hot spot of domestic crimes was distributed across southern and western parts of the city. Whereas, post-COVID-19 onset, the hot spot of non-domestic crimes was reduced across all the clusters whereas reported domestic crimes emerged northern and southern parts of the city. This observation implied the need to review innovative risk management strategies in response to the change in crime types.

3.3 Reasoning High Crime Rates

3.3.1 Influencing Factors from OLS Regression

The global OLS linear regression model was run for Period A and Period B respectively, and the results are shown in the following Tables 3 and 4.

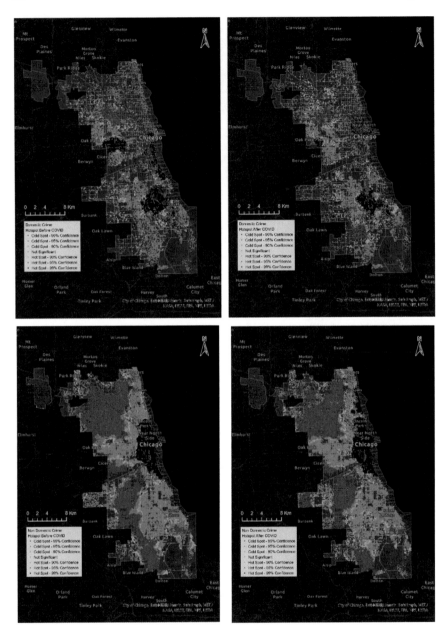

Fig. 8 Hotspots for non-domestic crimes (Top) and Domestic Crimes (Bottim) for pre-covid (Left) and post Covid (Right) in Chicago

Table 3 Summary of OLS regression—period A (pre COVID-19)

Variable	Coefficient	StdError	Probability	Robust_Pr	VIF
Intercept	−178.074219	92.366254	0.007*	0.006*	
Below Poverty Lv	12.696221	2.321613	0.001*	0.000*	2.759914
Crowded Housing	−4.858970	6.467990	0.005*	0.000*	2.167790
Per Capita Income	6.8237235	1.8425227	0.000*	0.034*	2.940431
Unemployment	12.073328	4.269899	0.006*	0.004*	3.492500
population density	−0.103025	0.085656	0.033	0.077	1.596234

Adjusted R-Squared:0.622
Koenker (BP) Statistic: 13.427*
*Indicates results are insignificant
Akaike's Information Criterion (AICc): 988.979

Table 4 Summary of OLS regression—period B (post COVID-19)

Variable	Coefficient	StdError	Probability	Robust_Pr	VIF
Intercept	−95.602	63.916	0.040	0.084	
Below poverty Lv	9.741	1.607	0.000*	0.000*	2.759914
Crowded housing	−6.861	4.476	0.129	0.082	2.167790
Per capita income	3.3923	1.2750	0.009*	0.052	2.940431
Unemployment	11.036	2.955	0.000*	0.001*	3.492500
Population density	−0.060	0.060	0.010	0.108	1.596234

Adjusted R-Squared:0.718
Koenker (BP) Statistic: 14.660*
*Indicates results are insignificant
Akaike's Information Criterion (AICc): 932.276

VIF values of all explanatory variables were less than 7.5, which meant there was no multicollinearity between variables and no redundancy in explanatory variables. Hence all explanatory variables could be retained. It was also seen in Table 3 that adjusted R-Squared was 0.622, which meant the independent variable could explain 62.2% of the crime distribution pre-COVID-19. Whereas in Table 4, the adjusted R-Squared was 0.718, that meant that the independent variable could explain 71.8% of the crime distribution post-COVID-19.

The adjusted R-Squared are both greater than 0.5, so the OLS model performed well.

To ensure that model residuals were free from spatial clustering of over and under predictions, spatial autocorrelation analysis on the standardized residual of OLS model results was carried out (Fig. 9d. The results are shown in Table 5).

Given the z-score of 1.530961, Since the z-scores were all less than 1.65 and P-values were not significant, the pattern did not appear to be significantly different than random, standardized residuals. Both periods did not exhibit spatial autocorrelation.

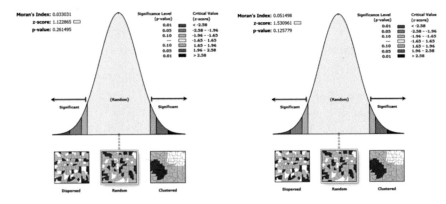

Fig. 9 10 Std residual report for period A (left) and period B (Right)

Table 5 Spatial autocorrelation od residuals

	StdResid for period A	StdResid for period B
Moran's Index	0.033031	0.051498
Variance	0.001692	0.001784
z-score	1.122865	1.530961
p-value	0.261495	0.125779

3.3.2 Interpretation

In Chicago, the population density factor had a negative correlation with crime rate. This meant that crime rate was lower in communities with relatively higher population density, which was different from the general perception that higher the population density, the more likely it is to have frequent crimes. Before the outbreak of COVID-19, the crime rate in Chicago increased by 0.103 for every 1 person/sq km decrease in population density, but the impact of population decreased after COVID-19 (the absolute value of the coefficient decreased to 0.06).

The economic conditions of community residents included two variables: 'Below Poverty Level' and 'Per Capita Income'. The below poverty level had a positive correlation with the crime rate in the community, which means the high poverty level leads to the relatively high crime rate. Per capita income has a similar positive correlation, which means that the rise in per capita income has led to the rise in crime. Although the analysis of these two variables were contradicting, it was inferred that in areas where both per capita income and poverty levels are high, it referred to the huge gap between the rich and the poor within the community. The model result found that these communities—with high per capita income and high poverty—had reported higher crime rates. Their coefficients fell during the COVID-19, suggesting that their influence on crime rates had declined.

The unemployment rate also positively correlates with the crime rate. This essentially means that the increase in unemployment rate led to increase in crime rate, which is consistent with the general perception.

The coefficient of crowded housing was negative, indicating that places with relatively high levels of crowded housing have relatively low crime rates. It was noteworthy that the absolute value of its coefficient was the only one that increased during the period of COVID-19. The effect of crowded housing on crime rates increased during the period of COVID-19 and this may have something to do with the fact that the government urged people not to go to crowded places.

4 Discussion and Conclusion

This study examined the changes in crime trends before and after COVID-19 in Chicago and evaluated how community characteristics influenced these crime changes.

Temporal and spatial distribution of cases of crime were presented in three ways: overall change of crime, domestic crime, and motor theft change. In general, crime hotspots clustered predominantly around urban districts before COVID-19, whereas more clusters emerged since the onset of the pandemic around the western and southern parts of the city. The coefficients in the results of OLS models also reflected the influence of the following explanatory variables in areas with higher reported crime rate post-COVID-19:

- In communities with lower population density, which differed from the usual perception that crime frequency increased according to the population density;
- In communities which reported higher income per capita and higher poverty level concurrently, as the opportunity and temptation are perceptively higher;
- In communities with higher unemployment rate; and lastly
- In communities where crowded housing are more prevalent.

It is noteworthy that domestic crime increased significantly during COVID-19 period and most hotspots appeared in southern parts of Chicago, which may be the main contribution of crime trends in the pandemic. However, this study was unable to establish the attributing factor(s) at this point due to the paucity of data. This finding detailed with concerns about pandemic-related domestic abuse as lockdowns and quarantines can trap vulnerable people like women and children, physically and mentally challenged among others who could be forced to stay with abusive people. This specific finding was also consistent with the OLS finding that crime rate corresponded positively in areas with crowded housing. When a victim was required to stay in a home without access to the usual outlets that help to reduce tension, such as time apart when at work, opportunities to visit friends or family or a private place to reach out for help—the opportunity for violence naturally increases. Stress stemming from the pandemic itself probably contributed as well.

Many shelters and safe homes have had to move people into hotels to comply with COVID-19 guidelines, so space for domestic abuse victims can be limited. But help could still be made available through hotlines and other domestic violence programmes. Even if leaving home and going to a shelter is not what a survivor wants to do, the staff at safe home programs could provide support, help to develop a plan to better keep oneself and children safe, as well as to connect to valuable resources in the community [10].

On the other hand, the changes in motor theft trends went down initially when the 'Stay at Home Order' was executed and went up the following month there is likely a perception of reduced public and police presence and that their probability of being caught is also reduced. Study conducted by [5] also explained this change of trend to be due to crime substitution—that is "the reduction in the probability of finding a victim for other crimes shifts the supply of time spent searching for a vehicle to steal". The increase in motor theft shows that some criminals seem quite rational in the choice of target. This specific finding is consistent with the OLS finding that crime rate corresponded positively in areas where the income per capita and poverty level are higher as areas with higher income per capita are likely to have more vehicles. As such, it might be wise to have more police patrols in these areas.

While the police were the most visible and maybe salient thing that society does to try and achieve public safety, there was also a need to invest in social safety net and programmes to keep children and youth safe and not just sitting around when schools are closed down during this period. This is specifically relevant in areas where unemployment rates are high and crowded housing is prevalent.

These current crime changes are the result of the movement restrictions that have been placed on people and places. While the criminal justice caseload is reduced, the real risk of crime may not have declined. As noted above, if the decline in reported crime was largely explained by the reduction in exposure to crime, the real risk of crime may have fallen very little.

As such, this study was conducted with the intention that these initial findings of the pandemic effects on crime will help inform those decisions. Future research could consider the association between crime rates with social and health data to ensure adequate investment in social safety net and programmes to strengthen social resilience.

References

1. ABC News (2020) Timeline: how coronavirus got started. https://abcnews.go.com/Health/timeline-coronavirus-started/story?id=69435165. Accessed 23 Oct 2020
2. IIEP (2020) Institute for international economic policy. https://iiep.gwu.edu/category/past-events/. Accessed 23 Oct 2020
3. WebMD (2020) Study finds rise in domestic violence during COVID. https://www.webmd.com/lung/news/20200818/radiology-study-suggests-horrifying-rise-in-domestic-violence-during-pandemic#1. Accessed October 23, 2020

4. Bird L, Bish A, Gastelum Felix S, Gastrow P, Mora Gomez M, Kemp W, Kimani J, Meegan-Vickers J, Mejdini F, Micallef M, Tagziria L, Tennant I, Rademeyer J, Reitano T, Scaturro R, Sellar J, Shaw M, Stanyard J (2020) Policy brief: the impact of a pandemic on organized crime - crime and contagion. https://globalinitiative.net/wp-content/uploads/2020/03/CovidPB1rev. 04.04.v1.pdf [i] . Accessed 23 Oct 2020
5. McDonald JF, Balkin S (2020) The COVID-19 and the decline in crime. SSRN Electron J. https://doi.org/10.2139/ssrn.3567500
6. Campedelli GM, Aziani A, Favarin S (2020) Exploring the effects of COVID-19 containment policies on crime: an empirical analysis of the short-term aftermath in Los Angeles. https://doi.org/10.31219/osf.io/gcpq8
7. Payne JL, Morgan A (2020) COVID-19 and violent crime: a comparison of recorded offence rates and dynamic forecasts (ARIMA) for March 2020 in Queensland, Australia. SocArXiv g4kh7, Center for open science
8. NYT (2020) The pandemic has hindered many of the best ideas for reducing violence. https://www.nytimes.com/interactive/2020/10/06/upshot/crime-pandemic-cities.html. Accessed 27 Oct 2020
9. Halford E, Dixon A, Farrell G, Malleson N, Tilley N (2020) Crime and coronavirus: social distancing, lockdown, and the mobility elasticity of crime. Crime Sci 9(1):11. https://doi.org/10.1186/s40163-020-00121-w
10. Leslie E, Wilson R (2020) Sheltering in place and domestic violence: evidence from calls for service during COVID-19. SSRN Electron J. https://doi.org/10.2139/ssrn.3600646
11. Piquero AR, Riddell JR, Bishopp SA, Narvey C, Reid JA, Piquero NL (2020) Staying home, staying safe? A short-term analysis of COVID-19 on dallas domestic violence. Am J Crim Justice 45(4):601–635. https://doi.org/10.1007/s12103-020-09531-7
12. Abrams DS (2020) COVID and crime: an early empirical look. SSRN Electron J. https://doi.org/10.2139/ssrn.3674032
13. WSJ (2020). Coronavirus pandemic changes policing, including fewer arrests. https://www.wsj.com/articles/coronavirus-pandemic-changes-policing-including-fewer-arrests-11585301402. Accessed 26 Oct 2020
14. LUC (2020) Covid-19's effect on Chicago crime trends. https://www.covidcrime-chicago.com/. Accessed 27 Oct 2020
15. Ester M, Kriegel HP, Sander J, Xu X (1996) A density-based algorithm for discovering clusters in large spatial databases with noise. Kdd 96(34):226–231
16. Chicago Stay At Home FAQ. (2020) https://www.chicago.gov/content/dam/city/depts/mayor/Press Room/Press Releases/2020/March/FAQsStayAtHomeOrder.pdf. Accessed 30 Oct 2020
17. Bachman D (2020) The economic impact of COVID-19 | deloitte insights. (2020) https://www2.deloitte.com/us/en/insights/economy/covid-19/economic-impact-covid-19.html. Accessed 30 Oct 2020

Geospatial Analysis of Grab Trips in Singapore

Huang Fengjue, Ji Xin, Zhu Wenzhe, and Hu Guanxian

Abstract It has been observed that more and more young people choose to use online car-hailing as the preferred travel mode. Hence a study of the trip patterns of online car-hailing can reveal a lot on the travel needs and patterns of the young generation. This study explores the characteristics of Grab trips, which is the major car hailing service provide in Singapore, order points in Singapore, by modelling them geospatially using their GPS tracks and points of interests in Singapore. The spatial analysis evaluates various regression models, namely the Ordinary Least Squares (OLS), Geographically Weighted Regression (GWR) and Multiscale Geographically Weighted Regression (MGWR). The MGWR results were found to be far superior to OLS and GWR. It was found that points of interest lie retail shopping outlets, clusters of young population, taxi pick points and public transport facilities impact the trip start and end points of car hailing services. Grab ride end points are primarily located at interests such as the city centre, airport areas, retail outlets and industrial areas which displays a spatial heterogeneity.

1 Introduction

Travel information of online car-hailing service providers like Grab can be used to reveal the needs and travel patterns of service consumers who are especially the young generation. In this study, the characteristics of Grab order points in Singapore are explored to determine relationships between the locations where the services are ordered with respect to the points of interest (POI) where the commuter alights. Regression models were used to examine the correlation of the chosen variables to understand their relationship. Spatial correlation methods were used to map the geographical distribution of these variables to identify patterns which can significantly influence the starting points of car hailing services. Grab ride trajectory data

H. Fengjue · J. Xin · Z. Wenzhe (✉) · H. Guanxian
Arts Link, National University of Singapore, 03-01, Block AS2, Singapore 117570, Singapore
e-mail: zhu_wenzhe102@163.com

© The Author(s), under exclusive license to Springer Nature Singapore Pte Ltd. 2022   155
S. N. Kundu (ed.), *Geospatial Data Analytics and Urban Applications*,
Advances in 21st Century Human Settlements,
https://doi.org/10.1007/978-981-16-7649-9_9

was used for spatial analysis along with other spatial data in Singapore like points of interest and land use categories.

1.1 Study Area

Singapore is located at 1°18' N and 103°51' E, with an area of 719.1 sq. km. In addition to the main Island (which accounts for 88.5% of the total area), Singapore also includes 63 surrounding islands. The terrain of Singapore is gentle, the western and central areas are composed of 400 hilly areas, most of which are covered by woods, and the eastern and coastal areas are plains. As of 2016, there are approximately 4 million citizens and permanent residents. Singapore is divided into five regions, including Central Region, East Community, North Region, North-East Region and West Region.

The Financial district of Singapore is located in the south-central part of the main island (Fig. 1). The only passenger airport, the Changi airport, is located at the east end of Singapore. The airport covers an area of over 13 sq. km and is an important hub for transport with the adjoining Changi Industrial park. The business district and the Changi region are the two regions where most people commute to and from on a regular basis.

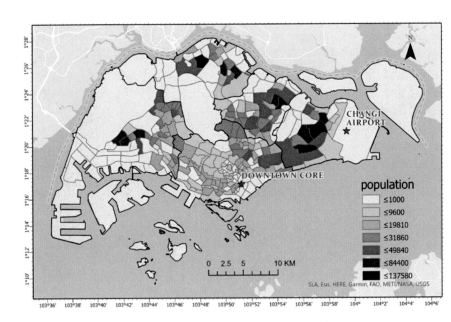

Fig. 1 The study area

1.2 Objective and Scope

In this study, multiple datasets were geospatially analysed. These include GPS based locational data for each trip, points of interest, population density and their distribution. The analysis was aimed to achieve the below objectives:

- Explore spatial–temporal distribution patterns Grab taxi orders.
- Use and compare several regression algorithms (OLS, GWR and MGWR) to evaluate results.
- To discourse and reason the starting points of taxi trips through spatial visualization and proximity analysis.

2 Data Used

2.1 Grab Posisi

Grab-Posisi, an extensive real-life GPS trajectory dataset in Singapore, was used to analyse the spatial–temporal distribution [2]. The data, hosted and shared by Grab Taxi Holdings, has attributed described in Table 1.

The data was wrangled to suit the purpose of the study. As the prime objective was to analyze the demand distribution for taxis, the first GPS location for each trip trajectory was extracted as it was regarding as the starting point for each journey. The "Ping Time Stamp", which was in UTC time, was converted to local Singapore time for temporal analysis. Three new attributes were extracted from the "Start Date and Time" attribute to separate fields for date and time too. Table 2 presents a sample of cleaned data for Grab Posisi which was used for the study.

The kernel Density of Grab Taxi starting points (Fig. 2) displays several clusters. The prime ones are located at Changi Airport and at the Central Business District (CBD). Other significant ones are the Changi Industrial Park (adjoining Changi Airport), Woodland Industrial Estate (in the north), NTU and Boon Lay Industrial Parks (in the west) and Yishun region (north-west). There are smaller clusters in

Table 1 Attributes of grab posisi data

Attributes	Data Type	Remark
Trajectory_ID	String	Identifier for the trajectory
Latitude	Float	WGS84
Longitude	Float	WGS84
Timestamp	Date	UTC
Accuracy Level	Float	Circle radius, in meter
Bearing	Float	Degrees relative to true north
Speed	Float	In meters/second

Table 2 Cleaned timestamp data

start date & time	Trajetory_ID	Ping time stamp	Latitude	Longitude	Speed	Bearing	Day of the week
2019–04-09 11:02:51	35,799	1,554,778,971	1.341502	103.709	12.76	145	2
2019–04-13 14:55:16	35,852	1,555,138,516	1.351276	103.7396	13.52864	178	6
2019–04-13 17:09:46	35,855	1,555,146,586	1.343412	103.8514	14.92415	278	6
2019–04-18 18:31:48	35,864	1,555,583,508	1.425584	103.8566	15.12738	125	4
2019–04-18 20:05:24	35,885	1,555,589,124	1.312283	103.8772	0	260	4

Fig. 2 Kernel density of taxi ride starting points

Punggol and along the east coast and west coast belts where the density of population is high.

Kernel density of trip starting points for weekdays and weekends (Fig. 3) were very similar although the volume of trips during weekends were lower (Fig. 4 right).

However, the diurnal distribution of taxi orders displays huge fluctuations (Fig. 4). The taxi orders increase significantly between 7 and 10 am and between 4 and 8 pm. These time windows correspond to people commuting to and from work. time periods

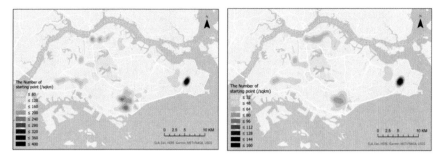

Fig. 3 Kernel density for weekdays (left) and weekends (right)

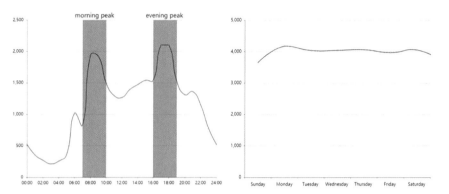

Fig. 4 Daily (left) and weekly (right) variation of Grab Orders

correspond to morning and evening peaks. This shows that there is a great demand for taxis in the morning and evening peaks (Fig. 4 left).

The order density patterns at morning peak period and evening peak period were very different (Fig. 5). In the morning peak period, more taxi starting points are concentrated in the north, west and east (airport) of Singapore, while during the

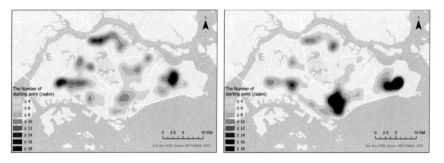

Fig. 5 Taxi starting points in the morning (left) and evenings (right)

Table 3 Data categories for POI data

Category	POI class
Retail	Supermarket, convenience, shopping mall, beauty shop
Entertainment	Bar, cinema, nightclub, pub
Catering	Restaurant, fast food, food court, cafe
Accommodation	Motel, hotel
Park	Park, attractions, theme park, zoo
Public traffic	Railway station, bus stop
Healthcare	Hospital, dentist
Education	University, school
Art	Bookshop, artwork, library, theatre
Finance	Bank

evening they are more concentrated in the downtown area around the Central Business District (CBD). The airport was an outlier as it showed similar density at both peak hours however the evening density was higher owning to people landing into Singapore and trying to get home.

2.2 Point of Interest (POI) Dataset

POI data was used to study the relationship between the taxi order distribution and the location of entertainment or other facilities. The POI data was sourced from OpenStreetMap (https://www.openstreetmap.org/). This project categorised POIs into groups to understand the relationship with trip trajectories (Table 3).

2.3 Resident Population Data

The data of the resident population was acquired from Data.gov.sg (https://data.gov.sg/). The resident population dataset contained the attributes of different population by subzone, age and sex. As the likelihood of the young generation of using Grab taxis was expected to be high, the proportion of young population from total population was used as a separate attribute.

3 Methodology

This study utilised an analytics workflow shown in Fig. 6, which was used for the

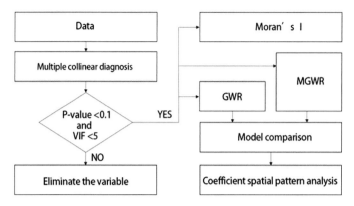

Fig. 6 The analytical workflow

Spatio-temporal distribution of taxi travel in Chengdu [3]. The experimental data was subjected to multiple collinearity analysis to isolate and eliminate the insignificant variable. This was done by using stepwise regression and least-squares regression methods. Next, the spatial autocorrelation of the remaining variables was performed using Moran's I methods. Finally, the classical Geographic Weighted Regression model and the Multi-scale Geographic Weighted Regression model was used to fit a regression model on each variable. The relevant results were then visualised and discussed.

3.1 Units of Spatial Analysis

To quantify the relative importance of geographic, social and economic factors with respect to Grab order location, a spatial unit of analysis was defined. Fishnets with an area of 1 sq. km. was used to summarise the attributes. Counts of each category of POI were summarised for this fishnet cell for aggregation. The use of fishnets helped standardise the study by creating discrete units for continuous spatial data. Without this spatial autocorrelation and cluster analysis could not have been applied.

3.2 Multiple Collinear Diagnosis and Global Regression

All attributes are likely to have some correlation although they are treated as independent variables. Therefore, it was essential to reduce the number of variables by removing the redundant ones. In this study, a stepwise regression model was used to diagnose multicollinearity and variables with a variance inflation factor (VIF) greater than 7.5 were eliminated. Least Squares Estimation (OLS) was performed on samples to screen statistically significant factors.

3.3   Spatial Autocorrelation

The global model based on least-squares estimates assumes that the effects of explanatory variables are spatially homogeneous. However, due to the heterogeneity in spatial distribution of variables, it was likely to have some spatial non-stationarity. Spatial autocorrelation helps understand the similarity between variables and the same variables in nearby cells [4]. The spatial distribution of residuals in OLS was calculated to check the independence of variables. If the residuals are clustering, then the variables are likely to be spatially dependent. Moran's I was used to explore whether the distribution order and the spatial distribution of residuals are spatially correlated. If yes, geographically weighted regression (GWR) was performed.

3.4   Geographical Weighted Regression(GWR)

Geographically weighted regression (GWR) explores the spatial changes and related driving factors of this study at a certain scale by establishing a local regression equation at each point in the spatial range. This was used to predict future results too as it took local effects of spatial objects into account and has the advantage of being more accurate. It is generally widely used to study spatial non-stationarity. The calculation formula of this model is formula (1).

$$y_i = \sum_r x_{ir}\beta_r(u_i, v_i) + \varepsilon_i \tag{1}$$

where (u_i, v_i) is the coordinate of the i point.

3.5   Multi-scale Geographic Weighted Regression (MGWR)

Fotheringham [1] proposed multi-scale geographic weighted regression (MGWR) based on a generalized additive model. This method allows each covariate to have a different spatial smoothing level, which used its optimal bandwidth for regression for each independent variable. It personalized each spatial process's spatial scale, resulting in closer to reality and more explanatory the spatial process model of force (Fig. 7). The calculation formula of this model is formula (2).

$$y_i = \sum_{j=1}^k \beta_{bwj}(u_i, v_i)x_{ij} + \varepsilon_i \tag{2}$$

where β_{bwj} refers to the bandwidth used by the regression coefficient of the j variable.

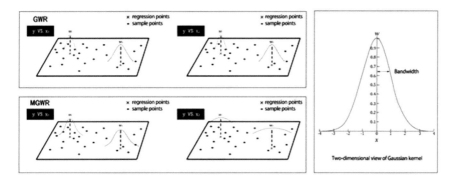

Fig. 7 Schematic diagram of GWR and MGWR models

4 Results

4.1 Regression Analysis

From the results of OLS regression (Tables 4 and 5), we get a statistically significant Koenker (BP) statistic. So, the Robust Probability is preferred to determine the significant variables. We find the variable "RETAIL" (number of retail facilities including supermarkets, convenience stores, malls and beauty shops), "YOUNG POPULATION" (population of young people), "TOTAL POPULATION" (total population), "TAXI POINT" (number of the location where people get in or get off the taxi cars), and "PUBLIC TRAFFIC" (traffic facilities such as car parks, bus stops, airport) and the significant ones for the study. However, the high VIF values of variable "YOUNG POPULATION" and "TOTAL POPULATION" indicate redundancy. Hence the latter was eliminated. The Adjusted R-Squared value was 0.5735. Subsequently it was checked whether the residuals showed normal distribution by using Spatial Autocorrelation (Moran's I) tool.

The Moran's I test for spatial autocorrelation yielded a value of 0.447, indicating strong spatial heterogeneity. This was consistent with the OLS regression results. The Local Indicator for Spatial Autocorrelation (LISA) analysis and map of significance (Fig. 8) displayed a high-high distribution around the city centre and Changi airport. Elsewhere the high-high distribution was scattered e.g., the Seletar region in the East and NTU at west of Singapore. High-high distribution is indicative of high number of trips. Likewise, the low-low regions were primarily distributed in the southwest and north of Singapore where industrial parks and nature reserves are located where the number of taxi trips were expected to be low in the temporal window. The map of significance for these regions were well within the 90% confidence, which indicates that the results are reliable.

Owing to the spatial heterogeneity of observed variables, the local regression model was more reliable for this analysis.

Table 4 Summary of OLS regression

Variable	Coef	Std error	T stats	Prob	Robust SE	Robust T	Robust Prob	VIF
INTERCEPT	3.9511	1.5052	2.625	0.0087*	2.0608	1.9173	0.0553	–
ENTERTAINMENT	−3.077	0.787	−3.9095	0.0001*	1.9429	−1.5837	0.1135	7.3938
HEALTHCARE	−9.0686	2.4978	−3.6307	0.0003*	5.377	−1.6865	0.0919	1.6675
RETAIL	4.1649	0.821	5.0733	0.0000*	1.8231	2.2845	0.0224*	7.1353
PARK	1.0318	0.6217	1.6596	0.0972	0.7576	1.362	0.1734	1.1682
ACCOMMODATION	0.7732	0.5536	1.3967	0.1627	1.0009	0.7725	0.4399	2.3639
ART	−1.1171	0.5581	−2.0015	0.0455*	1.3399	−0.8337	0.4045	3.0857
CATERING	0.1804	0.1588	1.1365	0.2559	0.4441	0.4062	0.6846	15.6594
YOUNG POPULATION	0.0117	0.0017	7.0857	0.0000*	0.0045	2.6272	0.0087*	251.1987
TOTAL POPULATION	−0.0058	0.0009	−6.2273	0.0000*	0.0025	−2.3669	0.0180*	252.6779
TAXI POINT	9.4076	1.0918	8.6162	0.0000*	2.5289	3.7201	0.0002*	4.6677
BANK	−0.0114	1.0336	−0.011	0.9912	2.0568	−0.0055	0.9956	2.1422
EDUCATION	−6.8574	2.5812	−2.6566	0.0080*	6.3466	−1.0805	0.2801	1.4531
PUBLIC TRAFFIC	2.8592	0.1831	15.620	0.0000*	0.2647	10.8005	0.0000*	2.6403
DISTANCE	−0.0001	0.0001	−2.1601	0.0309*	0.0001	−1.7291	0.084	1.0486

* means the coefficient is statistically significant

Table 5 OLS diagnostics

Diag_name	Diag_value
AIC	18,025.886260
AICc	18,026.176703
R2	0.576625
AdjR2	0.573463
F−Stat	182.406667
F−Prob	0.000000*
Wald	758.992817
Wald−Prob	0.000000*
K(BP)	56.518154
K(BP)−Prob	0.000000*
JB	752,937.883615
JB−Prob	0.000000*
Sigma2	804.857984

* means the coefficient is statistically significant

Fig. 8 LISA (left) and map of significance (right)

4.2 GWR and MGWR Results

Table 6 lists the comparison between GWR and MGWR. The residual of MGWR do not show an obvious large area of spatial aggregation and therefore, it can be considered that the residuals were randomly distributed in space, indicating a good fit (Fig. 9 left).

The spatial distribution of intercept coefficients (Fig. 10 top left) represented the sum of all the objective factors affecting the dependent variables except the independent variables used. These intercepts can be simplified to the effect of geography on the distribution of Grab trips. A strip stretching from the CBD to Changi International Airport in the southeast forms the hot zone that impacts the Grab trips' distribution. This means that the geographical location of this part has a strong positive effect on Grab trajectories.

Table 6 Comparison of GWR and MGWR

Diagnostic information	GWR	MGWR
Residual sum of squares	200.534	181.454
Effective number of parameters (trace(S))	55.804	99.6
Sigma estimate	0.64	0.638
Log-likelihood	−500.876	−473.631
AICc	1194.565	1128.84
R2	0.632	0.667
Adj. R2	0.583	0.592

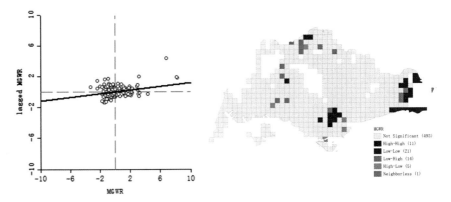

Fig. 9 Moran's I of residual GWR (left) and its LISA (right)

Figure 10 (top right) displays the coefficients of retail facilities contributing to the Grab travels. In urban centers, retail outlets are relatively saturated and thus has a low marginal effect on Grab orders. This essentially means that more retail stores do not generate more travel orders. Also there appears to be an effect spillover, resulting in a negative correlation between retail and travel in some areas. In contrast, retail outlets in industrial areas has a high marginal effect on Grab travels, and the distribution and setting of retail outlets can lead to more Grab orders.

Figure 10 (bottom left) displays the impact of the young population on Grab trips. The influence is high although the trend decreases from the west to the east. In the west of Singapore, the young population is a major impetus to Grab travels, whereas in the eastern island, the impetus could be because of other factors.

Figure 10 (bottom right) shows that the number of the taxi hailing points positively affect the Grab orders, but the contrast is too low indicating that the influence is fairly even across the island. The likely explanation for the slightly decreasing influence from northwest to southeast is that public transport in Singapore is more developed in southeast and therefore this part do not generate more Grab orders.

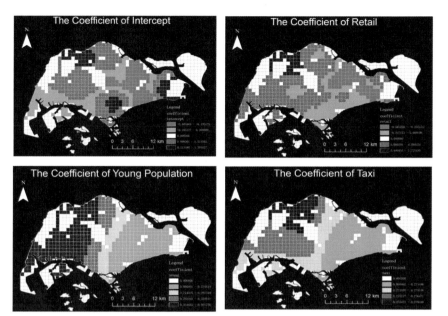

Fig. 10 Spatial patterns of coefficients

So far as public traffic stops are concerned, more public traffic stops are indicative of more well-developed infrastructure and residential land use. Hence the contribution to grab orders from such areas are higher than the west of Singapore. It was observed that the impacts on Grab orders by public traffic facilities and taxi points are spatially contrasting as they both complement each other. Grab start points therefore originate from locations where there the density of bus stops and mass rapid transit (MRT) stops are less.

5   Conclusion

On a temporal context, Grab orders show clear peaks during the day although weekend and weekdays were very similar. Spatially they show clear clustering. Hotspots are at Industrial parks, Retail outlets, Airport and the CBD. Cold spots are primarily at places where there are natural parks and forests. The MGWR model was found to be more useful than OLS and GWR because of its ability to capture different scales of influence for different variables. This also helped reduce bias in the analysis.

The distribution of retail, young population, taxi points and public transport facilities influence a lot on the Grab orders making them complementary and hence heterogeneous. significantly on Grab orders and all of them show a spatial hetero-geneity. This is important for public administration to plan for the future on improving connectivity of public transport. This also calls for an integrated analysis of Grab demands vis-à-vis traffic congestion to optimize travel on a daily basis. For car hailing service providers like Grab, in addition to conventional operation, the spatial distribution of starting points can be factored to station available cars at different time of the day and week. As such car hailing services complement public transport other factors can be considered to enhance business and services by incorporating travel needs and patterns of individuals. Targeted marking to Grab subscribers can also be enhanced.

5.1 Limitations

There are four significant variables in OLS regression: retail, young population, taxi points and public transport facilities. More factors could potentially be included for a targeted study. The grid size of 1 sq. km can also be reduced to check is the study has a scale dependency. Weighted land use categories could also be helpful in enhancing the results. Despite elimination of several variables, the element of multicollinearity was not eliminated. The coefficients of the intercepts too do not collect enough explanatory variables exists among the variables selected in the project.

Despite these limitations, the current study is a demonstration of what can be achieved using spatial data which is being collected continuous for each trip. There are several perspectives to view this study: one from a public servant and the other from a busines perspective.

References

1. Fotheringham AS, Yang W, Kang W (2017) Multiscale geographically weighted regression (MGWR). Ann Am Assoc Geogr 107(6):1247–1265. https://doi.org/10.1080/24694452.2017. 1352480
2. Huang X, Yin Y, Lim S, Wang G, Hu B, Varadarajan J, Zheng S, Bulusu A, Zimmermann R (2019) Grab-posisi: an extensive real-life GPS trajectory dataset in southeast Asia. In: Proceedings of the 3rd ACM SIGSPATIAL international workshop on prediction of human mobility, pp 1–10. https://doi.org/10.1145/3356995.3364536

3. Li T, Jing P, Li L, Sun D, Yan W (2019) Revealing the varying impact of urban built environment on online car-hailing travel in spatio-temporal dimension: an exploratory analysis in Chengdu China. Sustainability 11(5):1336. https://doi.org/10.3390/su11051336
4. Moran PAP (1950) Notes on continuous stochastic phenomena. Biometrika 37(1/2):17. https://doi.org/10.2307/2332142

Ecological Vulnerability of Nyingchi, Tibet

Yan ChengCheng, Wang YunJie, Luo LaiWen, and Qi RuiYuan

Abstract The Alpine ecological region of Qinghai Tibet Plateau is undergoing significant changes endangering its vulnerability to ecology. Based on Pressure-State-Response (PSR) model, six indexes were are analyzed to evaluate the ecological vulnerability of Nyingchi city in Tibet. The Spatio-temporal ecological vulnerability of Nyingchi city from 2000 to 2015 was quantified based on spatial modeling techniques and observed changes were discussed in the context of space and time. The Ecological Vulnerability Standardized Index (EVSI), which was obtained by accounting different vulnerability levels yielded that the overall vulnerability of Nyingchi is relatively high in the north and gradually decreasing towards the south. The changes observed correlated well with local topography, climate, disaster volume and water conservancy construction. As a result, five levels of ecologically fragility could be defined for Nyingchi city, which shall provide a direction to efforts for ecological restoration and re-construction. From the perspective of ecological restoration and reconstruction planning of restoration and reconstruction can be addressed based on the severity of the vulnerability. Such studies on other vulnerable cities at a closer time periods can establish a continuous monitoring systems to detect changes in vulnerability. Automated systems with continuous feed of satellite data with dynamic Realtime processing can help measure and feed information to information dashboards which monitor ecological health of a region. These systems can ensure sustainability of our ecology.

1 Introduction

Ecosystem one of the basic component units of our planet, and it is important for the sustainable development of society [2]. In recent years, researchers have increasing focussed on sustainable development to address change of the global environment.

Y. ChengCheng (✉) · W. YunJie · L. LaiWen · Q. RuiYuan
National University of Singapore, 21 Lower Kent Ridge Rd, Singapore 119077, Singapore
e-mail: yancc@tsinghua-sh.cn

© The Author(s), under exclusive license to Springer Nature Singapore Pte Ltd. 2022
S. N. Kundu (ed.), *Geospatial Data Analytics and Urban Applications*,
Advances in 21st Century Human Settlements,
https://doi.org/10.1007/978-981-16-7649-9_10

This places ecological vulnerability analysis as one of the core study for sustainable practices in our society.

In this perspective, the Qinghai-Tibet Plateau, has been found to be one region which is undergoing significant ecological changes due to multiple factors which mostly anthropogenic ones and its domino effects. As more than 80% of the plateau area is above 4000 m above sea level, the temperature is significantly lower than other regions at the same latitude area, making this the "world's third pole". There is more precipitation in the southern and south-eastern edge of the Qinghai Tibet Plateau, while in the vast hinterland receives precipitation less than 200 mm. The history of soil development is short as most sediments are or glacial origin. 71.67% of the plateau is covered with alpine soil, which is characterized by weak corrosion resistance. The vegetation is mainly alpine meadow and grassland with a single structure; low temperature and water shortage make grassland productivity low and regeneration slow [4]. These factors make the region more sensitive to external disturbance and therefore the Qinghai Tibet Plateau prone to degradation with the slightest change in climate and anthropogenic activity.

Anthropogenic activities have impacted the stability of the region's ecosystem and has triggered local, short-term changes in the plateau, and is now the main cause of regional ecological degradation [4]. Overgrazing is the first cause of grassland degradation, and the overload rate in the northern Tibetan Plateau is 59.18% [7]. In addition, the exploitation of mineral resources and the development of biological and tourism resources have caused serious damage to the ecology through grassland degradation, disappearing wetlands, desertification etc.

Global climate change also affects the vulnerability of ecosystems [5]. Due to the extreme climate and harsh natural conditions in the Qinghai Tibet Plateau, the key environmental factors in the system are often in a critical state as natural resistance is poor. Therefore, it is natural that climate change would have an impact on the regional environment and could cause changes in the pattern, process and function of the plateau's ecosystem. Past observed evidence has confirmed that global climate change has led to the accelerated retreat of glaciers on the plateau, reducing the thickness and depth of the permafrost layer and adding to extent of wetlands. This has a slow but long-term impact on the ecosystems in alpine regions [5].

Nyingchi City is a typical area which is representative of Qinghai-Tibet Plateau. The city area has an average elevation of 3100 m with a forest coverage of about 46.09%. The natural resources of the city area are the wetlands and the forests and this makes the region suitable for this study of ecological vulnerability.

1.1 Objectives and Scope

For Nyingchi city region (Fig. 1), 6 indexes were chosen to build an ecological vulnerability model. The scope of this study could be explained primarily as the below two;

Fig. 1 Location map of Nyingchi City in Qinghai

- Determine the spatial distribution of ecological vulnerability in Nyingchi city area using PSR model
- Explore influence factors of ecological vulnerability and present a perspective on sustainable development.

The ecological vulnerability evaluation system was based on PSR model. The system was implemented, and results visualised in ArcGIS software. Spatial Principal Component Analysis (SPCA) was performed on several variables to choose the 6 most important ones which were used to evaluate the ecological vulnerability index and study the spatial distribution over the study area between 2000 and 2015. Using techniques like Hot Spot Analysis (Getis-Ord Gi* statistics), the weighted Ecological Vulnerability Index (EVI) was rendered, and the ecological fragility of each region presented. Change and trends of EVI from 2000 to 2015 were reported and critically discussed. Based on the change trends, 5 levels of ecologically fragile areas were identified which set the stage for suggestions on impactful restoration and construction drives.

1.2 Data

Six evaluation indexes (Table 1) for the region was used for the study. The data for these indexes for the year 2000 and 2015 was downloaded from http://www.resdc.cn.

Since the data was of a spatial extent covering the whole of China, it was processed to exclude the regions outside of Nyingchi City administrative boundary. The data was in form of gridded raster which varied in spatial resolution and hence they were

Table 1 Data category ad themes

Category	Theme
Terrain	DEM
Earth's surface	NDVI
Meteorology	Annual mean temperature
Meteorology	Annual mean precipitation
Human Interference	Population density
Human Interference	GDP per capital
Nyingchi	Nyingchi administrative region

resampled to a sq km grid and reprojected to 1 km grid. The Geographical projection WGS84 was adopted for the overlay analysis.

2 Methodology

The workflow was adopted for incorporating the various methodologies and to analyse the data is provided in Fig. 2. It includes several processes elaborated as below.

2.1 Data Screening and Pre-processing

The root causes and influencing factors of ecological problems in the study area were identified and were used as a basis for evaluating vulnerability. Principal component analysis (PCA) was used to check degree of correlation between these factors to identify and remove redundant factors. Weights for each of the evaluation indexes

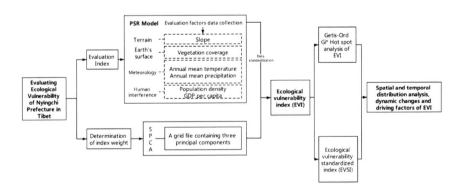

Fig. 2 Methodology and workflow

Table 2 Indexed and their ecological impact

Layer A	Layer B		Layer C		Positive or negative
	Name	Code	Name	Code	
Evaluation of ecological vulnerability in Nyingchi City	Terrain	B1	Slope	C11	+
	Surface	B2	Vegetation coverage	C21	−
	Meteorology	B3	Annual mean temperature	C31	−
			Annual mean precipitation	C32	−
	Human Disturbance	B4	Population density	C41	+
			Per capita GDP	C42	−

* " + " represents the positive relationship; "−" stands for inverse relationship
Slope data is selected as the representative index, and the computing process uses DEM data with a spatial resolution of 1 km as the source data, via Spatial Analyst module in ArcGIS 10.2 [6]

was determined objectively to avoid the effects of subjective arbitrariness and due to the large number and correlation of evaluation indexes [10]. In summary, this process helped identify the prime causative factors and their weights for evaluating ecological vulnerability.

Evaluation Indexes

The six indexes have conventionally been used in PSR models (e.g. [8]) for ecological vulnerability analysis. Their codification and impact on ecology is provided in Table 2.

Standardization

In the evaluation system, the dimensions and physical meanings of each evaluation factor are different. The data types, and spatial accuracy of different evaluation indicators are also different. Therefore, it is necessary to adopt a unified method to normalize and quantify them to eliminate the impact of discontinuity. The evaluation index system is continuous data for which the extreme value method was used to normalize the data. And all indexes of different dimensions and value ranges were unified to (0, 1) [8].

The relationship between each evaluation index and the ecological vulnerability can either be positive or negative. It is positive when an increase in the index results in an increase in ecological vulnerability. Else the relationship is negative. Slope and population density share a positive relationship with ecological vulnerability whereas vegetation coverage, annual average temperature, average annual rainfall, and per capita GDP had negative relationships. This relationship governed the range standardization used for the study.

For the index data of positive relationship, the normalization method is as follows:

$$P_i = \frac{X_i - X_{min}}{X_{max} - X_{min}} \qquad (1)$$

For the index data of negative relationship, the normalization method is as follows:

$$P_i = \frac{X_{max} - X_i}{X_{max} - X_{min}} \qquad (2)$$

where P_i is the standardized value of the ith index with the range of (0, 1), X_i is the actual value of the ith index of the year, Xmin is the minimum value of the actual value, and Xmax is the maximum value of the actual value. ArcGIS raster calculator function was used, and the normalised indexes are provide in Fig. 3.

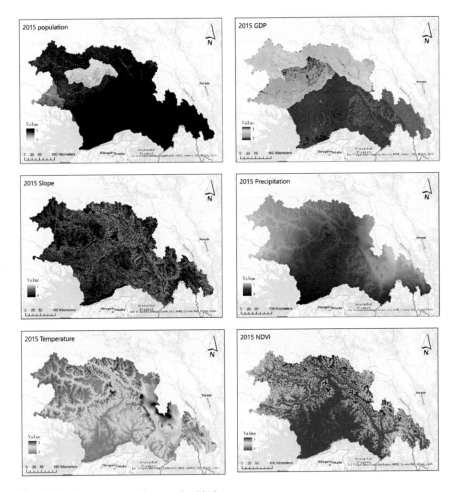

Fig. 3 The six standardised indexes for 2015

2.2 Analytical Methods

Spatial Principal Component Analysis (SPCA)

PCA transforms the above indexes into comprehensive evaluating index for ecological vulnerability assessment. It can compress the data set and convert the index data into a variety of representative comprehensive data [3]. PCA can be expressed as follows:

$$Y = n_1 x_1 + n_2 x_2 + \cdots + n_m x_m \tag{3}$$

where Y is the final component data, n is the component load, x is the actual value of the variable, and m is the total number of variables.

The spatial principal component analysis (SPCA) tool in ArcGIS was used to transform the data in the input band of multi-attribute space into a new multi-attribute space which rotates the axis relative to the original space. A set of raster bands were synthesized to generate a single band grid according to different contribution rates, and the specified principal components can be input Number. SPCA has some advantages over the conventional orthogonal function, as the data obtained are not any predetermined form, but are developed from data matrix to unique function, which can be used to explain the results.

Therefore, this study uses the above method for establishing the evaluation model of ecological vulnerability index (EVI) which outputs the principal components that affect the ecological vulnerability and conducts comprehensive treatment according to the correlation between the original indicators, so as to achieve the purpose of comprehensive analysis of ecological vulnerability by replacing more factors with a few common factors [11].

Ecological Vulnerability Index (EVI)

The Ecological Vulnerability Index is the combined value of multiple principal components and their corresponding weights, which can be expressed by the following formula [9]:

$$EVI = n_1 F_1 + n_2 F_2 + \cdots + n_m F_m \tag{4}$$

To quantitatively assess the variation trend of ecological vulnerability, the ecological vulnerability standardization index (EVSI) needs to be calculated from EVI. EVSI, which formed a comprehensive index of ecological vulnerability in Nyingchi city and is expressed by the below formula.

$$EVSI_i = \frac{EVI_i - EVI_{min}}{EVI_{max} - EVI_{min}} \tag{5}$$

According to the classification method of natural discontinuous points, the study area was divided into five levels of vulnerability index, and assigned values from low to high as micro, mild, moderate, moderate and extremely vulnerable [1].

Trend Analysis and Hotspot Analysis

The Hotspot analysis calculated the Getis-Ord Gi* statistics for each element in the dataset. From the z-score and p-value, the spatial clustering location of high-value or low-value elements was calculated as follows:

$$Gi^* = \frac{\sum_{j=1}^{n} w_{ij}x_j - \overline{X}\sum_{j=1}^{n} w_{ij}}{s\sqrt{\left[n\sum_{j=1}^{n} w_{ij}^2 - \left(\sum_{j=1}^{n} w_{ij}\right)^2\right]/(n-1)}} \tag{6}$$

$$\overline{X} = \frac{1}{n}\sum_{j=1}^{n} x_j \tag{7}$$

$$S = \sqrt{\left(\frac{1}{n}\sum_{j=1}^{n} x_j^2 - \overline{X}^2\right)} \tag{8}$$

where, G_i^* is the output statistics Z score; x_j is the EVI index change of spatial unit j; w_{ij} is the spatial weight between adjacent spatial unit i and j (Bai 2009).

3 Analysis and Results

3.1 Ecological Vulnerability

SPCA was performed on six evaluation indices to yield three resultant indexes whose Cumulative Contribution Rate was more than 90% (Table 3) for both 2000 and 2015.

According to the above method and formula, the Ecological Vulnerability Index (EVI) of Nyingchi city in 2000 and 2015 was be obtained by adding the product of the three principal components and their corresponding weights, as shown in the following formula:

$$EVI_i = \sum_{m=1}^{m}(W_i * A_i) \tag{9}$$

where, "i" represents the year of calculation, "W" represents the weight of the corresponding principal component, which was the normalized contribution rate, "A" represents the principal component vector, and "m" represents the number of principal components (m = 3). The calculation results are shown in Table 4.

Table 3 Statistics for principal component analysis

	PC	2000	2015
Eigen value	1	0.02437	0.03116
	2	0.00735	0.00789
	3	0.00523	0.00590
Percent of eigen values/%	1	60.5398	65.1166
	2	18.2701	16.4938
	3	12.9872	12.3382
Accumulative of eigen values/%	1	60.5398	65.1166
	2	78.8099	81.6104
	3	91.7971	93.9487
Weights	1	0.6595	0.6932
	2	0.1989	0.1755
	3	0.1416	0.1313

Table 4 EVI and EVSI for Nyingchi City in 2000 and 2015

Year	EVI (2000)	EVI (2015)	EVSI (2000)	EVSI (2015)	EVSI (SUM)
Minimum value	0.1488	0.0661			
Maximum value	1.1112	1.1284			
Mean value	0.5907	0.5975	45.9118	50.0178	47.9648
Standard deviation	0.2036	0.2395	21.1524	22.549	21.8507

A standardised value for Ecological Vulnerability Index (EVI) was achieved using formula (10) which ranged between 0 and 100.

$$EVSI_i = 100 \times \frac{EVI_i - EVI_{min}}{EVI_{max} - EVI_{min}} \qquad (10)$$

According to the results of the comprehensive analysis of score (Table 4), for the years 2000 and 2015, the mean value of Ecological Vulnerability Standardized Index (EVSI) was 45.9118 and 50.0178 respectively. Using natural breaks (Jenks method), the EVSI was classified into 5 categories: "Extremely Light", "Light", "Moderate", "Severe" and "Extremely Severe" (Tables 4 and 5).

3.2 Spatial Variability of EVSI

The EVI Ecological Vulnerability in Nyingchi city varied spatially. With a mean EVSI value of 47.9648 ± 21.8507, the whole region was found to be in a moderately fragile state. A general trend of spatial variability was seen in both 2000 and 2015

Table 5 Vulnerability classification based on EVSI

	Level of vulnerability	EVSI threshold partitioning
1	Extremely light	< 24
2	Light	24–40
3	Moderate	40–55
4	Severe	55–73
5	Extremely severe	> 73

(Fig. 4). EVSI was low in the south-central plains and high in the surrounding hilly regions I the north east and west.

In 2000, the extreme and severe ecological vulnerability of Nyingchi city tended to be mainly distributed in Gongbujiangda County, Bomi County, southwest of Lang County, northwest of Motuo County and north of Chayu County. The total area of this kind of region was 40,077.04 km², accounting for 34.93% of the total area of Nyingchi City. The moderately ecological vulnerability of Nyingchi city tended to be mainly distributed in Linzhi County, Milin County and southeast of Chayu County. The total area of this kind of region was 27,619.16 km², accounting for 27.41% of the total area of Nyingchi City. The extremely light and light vulnerability of Nyingchi city tended to be mainly distributed in south of Motuo County and southwest of Chayu County. The total area of this kind of region was 46,905.82 km², accounting for 37.66%.

In 2015, the extreme and severe ecological vulnerability of Nyingchi city tended to be mainly distributed in Lang County, Gongbujingda County, Bomi County, west of Linzhi County, west of Milin County, north of Chayu County and northwest of Muotuo County. The total area of this kind of region was 44,739.01 km², accounting for 38.99% of the total area of Nyingchi City. The moderate ecological vulnerability of Nyingchi city tended to be mainly distributed in east of Linzhi County, east of Milin County and southeast of Chayu County. The total area of this kind of region was 27,126.01 km², accounting for 23.64% of the total area of Nyingchi City. The extremely light and light vulnerability of Nyingchi city tended to be mainly distributed in south and east of Muotuo County and southwest of Chayu County. The total area of this kind of region was 42,874 km², accounting for 37.37% of the total area of Nyingchi City.

To sum up, from 2000 to 2015, the ecological vulnerability of Nyingchi city increased significantly, and the overall trend being southward. As can be seen from Table 6, the area with the largest proportion in Nyingchi city is light vulnerability, and its comprehensive area in the study area accounts for 23.9%. The second area is the moderately vulnerable area, which accounts for 23.86% of the comprehensive area in the study area. It is very similar to the lightly vulnerable area. Extremely light region was the region with the smallest proportion, and its comprehensive proportion in the study area was 15.23%. The comprehensive proportion of severe and extremely severe regions in the study area is 20.64% and 16.39%, respectively.

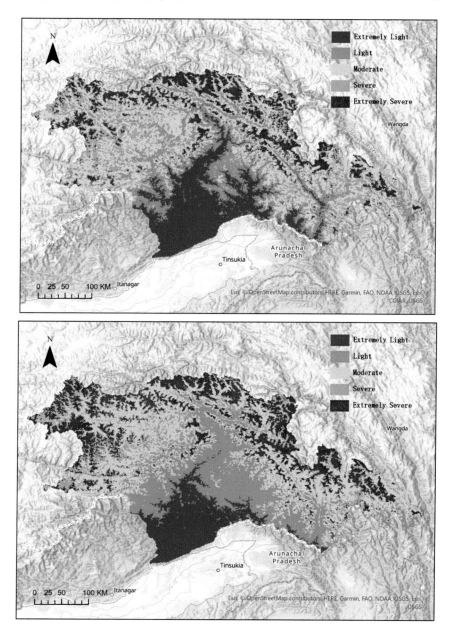

Fig. 4 EVSI in 2000 (Top) and 2015 (Bottom)

Table 6 Extent of each vulnerability class (area in sq km)

Year	Vulnerability class	2000	2015	Mean
Area (km²)	Extremely light	20,929.65	14,014.00	17,471.83
	Light	25,976.17	28,860.99	27,418.58
	Moderate	27,619.16	27,126.01	27,372.59
	Severe	24,918.32	22,435.94	23,677.13
	Extremely severe	15,158.73	22,303.07	18,730.90
Percentage (%) of Total Area	Extremely light	18.24	12.21	15.23
	Light	22.64	25.15	23.90
	Moderate	24.07	23.64	23.86
	Severe	21.72	19.55	20.64
	Extremely severe	13.33	19.45	16.39

3.3 EVSI Change Trends

In a period of 15 years, the proportion of areas with all degrees of vulnerability in Nyingchi city had a noticeable change (Fig. 5). Extremely light vulnerable areas decreased whereas extremely severe vulnerable areas increased significantly.

The proportion of extremely light vulnerable areas decreased from 18.24% to 12.21% whose rate of change was −33.06%. And the proportion of extremely severe areas increased from 13.33% to 19.45% whose rate of change was 45.91%. The moderately vulnerable areas had the least change from 24.07% to 23.64% whose rate of change was −1.79%. The proportion of light vulnerable areas increased from 22.64% to 25.15% whose rate of change was 11.09%. And the proportion of severe vulnerable areas decreased from 21.72% to 19.55% whose rate of change was − 9.99%.

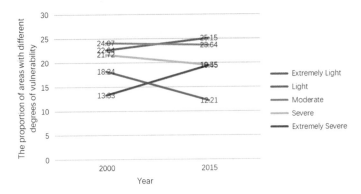

Fig. 5 Change Trend for each EVI class

3.4 Analysis of Hotspots

To identify hotspots and cold spots of ecological vulnerability in Nyingchi city, "Getis-Ord Gi*" statistics were generated using the weighted EVI data for 2000 and 2015 (Fig. 6). The cold spots appear to cluster in the south of the study area in both 2000 and 2015 indicating that this region underwent very little changes in EVSI. This finding is supported by the fact that this area is where the Medog country protected forest stock is location where the forest cover is around 80%. This is little touched by anthropogenic activity and the biodiversity of the region is high.

The hot spots mainly cluster in the west and north of the Nyingchi city. In 2000, Gongbujiangda County, Nyingchi County, Bomi County and Zayu County appear to be significant hot spots indicating very high ecological vulnerability. However, in 2015, some regions appear to have turned cold spots, especially in Nyingchi County and Zayu County, indicating a reduced ecological vulnerability than in year 2000. This could be the result of the sand prevention afforestation project implemented in year 2009 when 70 ecological safety barriers we constructed to preserve the ecosystem. In Zayu County, many National Nature Reserves were defined since 2002 to address the ecological problem in the area and these appear to have resulted in reducing the ecological vulnerability of the area.

In summary, the hot spots and cold spots both turn to be less significant from 2000 to 2015. The ecological vulnerability of the Nyingchi city seems to be improving where there are significant hotspots decrease evidently. This tells a lot about the past conservation efforts from the government of Nyingchi city.

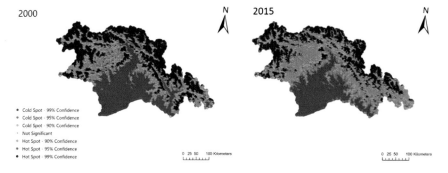

Fig. 6 EVI Hotspots (red) and Cold spots (blue) in 2000 (left) and 2015 (right)

4 Discussion

4.1 Role of Natural Factors

Natural environmental factors on ecological vulnerability in Nyingchi city were identified to be elevation, geological hazards, drought and climatic. According to the index system established in the study, slope and topographic relief are the most important influencing factors on ecological vulnerability.

Being located in the Himalayas, the study area has obvious vertical zonal distribution characteristics. In Nyingchi city, the highest elevations are to the north whereas the elevation of the south is lowers with a flat topography. Therefore, elevation is a defining factor for ecological vulnerability of Nyingchi city. Data from 2000 to 2015, clearly show that the northern area is relatively stable ecologically. However, additional factors, such as low temperature, low precipitation, heavy hail and snow, and strong wind in the high-elevation area, as well as the severe freeze–thaw erosion, influences the ecological vulnerability at these high elevations where vegetation is low, and population is dense.

The southeastern low-elevation region is a contrast as it has warm and humid airflow, abundant precipitation, high vegetation coverage, and little human influence. These factors reduce the ecological vulnerability. As half of the area of Nyingchi city is glacier plateau, most of the rivers, lakes and wetlands of Nyingchi city originate from the huge reservoir of snow mountain, which exerts a great influence on the local farmland irrigation, pasture breeding, water conservancy and power generation. Global warming, shrinking glaciers and melting snow have led to frequent disasters in Nyingchi city. These climate related disasters make the area more ecologically vulnerable.

From a climate perspective, the main causes of natural disasters in Nyingchi city, was driven by the arid and semi-arid monsoon climate. The temperate zone in Nyingchi is the main cause of drought, high winds and floods in the region. Additionally, such extreme events tend to remove the loos soil cover on the North bank of Nyingchi leading to reduction of vegetated cover. In winter and spring, the high wind and drought are usually synchronized, and the seasonal effect makes the water level of the Yarlung Zangbo River drop, leading to the exposure of a large area of the beach on both sides. This provides sufficient sand source for the large-scale activities of wind-sand.

4.2 Role of Anthropogenic Factors

The impact of human factors on ecological vulnerability can be divided into unreasonable use of land resources, infrastructure projects, water conservancy projects and the impact of agriculture and animal husbandry industry. Land use change, population rise, animal husbandry, landscape diversity and other factors add to the ecological vulnerability equation.

The overall trend shows that the highly vulnerable area in Nyingchi city is the urban area with the most intense human activities, and its single structure and function leads to its weak ability to resist external interference.

From the perspective of the impact of infrastructure projects on the ecological vulnerability change in Nyingchi city, the large amount and long construction time of infrastructure projects have affected the normal habitat of plateau animals. The infrastructure planning of Nyingchi city can be divided into three types. One type is point-like infrastructure, such as water source and sewage pond project in water supply and drainage system, substation of electric power project, dams, irrigation station, agricultural machinery well and reservoir in agriculture and animal husbandry water conservancy project. The second type is road engineering, water supply and drainage pipeline laying and other linear infrastructure; The third type is surface infrastructure, mainly including the transformation of market towns and residential areas of farming and animal husbandry villages. During the implementation of infrastructure projects, especially when earth is extracted from steep slopes, slope vegetation will be destroyed, and soil erosion will be easily caused. Therefore, the importance of afforestation and water conservancy construction to reduce the ecological vulnerability of Nyingchi city was obvious.

4.3   Ecological Vulnerability and Sustainable Development

Promoting the coordinated and sustainable development of economy, society and resources and ecology is the current goal and demand of industrial structure and production mode. Due to the great changes in population and economic development, the increase of population is also an important factor for the deterioration of the ecology. In addition, with the increase of population, the demand for means of production and consumption will increase, and the demand for resource development will intensify, which will further pose greater pressure and threat to the natural environment. With the development of urbanization, the influence of the increasing urban population on the local temperature will also lead to different degrees of urban heat island effect, which will further affect the local radiation, making the temperature of the ground and the lower atmosphere rise along with it. On the one hand, increased air temperature may improve regional evaporation, which is not conducive to vegetation growth; on the other hand, it extends vegetation growth cycle and improves precipitation and surface vegetation productivity. In addition, global warming affects the anti-interference ability of the ecosystem, and the impact on the fragile ecosystem will be more obvious, resulting in a greater human demand for and destruction of natural resources, which further increases the ecological vulnerability of Nyingchi city.

Finding alternatives to traditional energy and clean energy and implementation of sustainable utilization of resources engineering, can promote the green development, good circulation, low carbon development, formation of saving resources and protecting the environment of the industrial structure, mode of production, and turn from the source of ecology deterioration, create a good production and living environment.

5   Conclusion

In several past studies on ecology assessment, dynamic grid analysis was rarely carried out through remote sensing data and GIS technology. Most relied on economic statistics of administrative units or carried assessment for a time period. This study made use of remote sensing data, meteorological data and economic statistics data to establish four levels of the six evaluation indexes and was analysed spatially to evaluate the ecological vulnerability. This formed the very basis for discussing the spatial vulnerability and its spatial distribution for two temporal periods to establish change trends in the spatial context. The study identified five ecological vulnerability zones for the study area where different levels of ecological protection and restoration can be addressed. Such studies on other vulnerable cities at a closer time periods can establish a continuous monitoring systems to detect changes in vulnerability. Automated systems with continuous feed of satellite data with dynamic Realtime processing can help measure and feed information to information dashboards which monitor ecological health of a region. These systems can ensure sustainability of our ecology.

References

1. Bai Y, Ma H, Zhang B, Liang J, Li Z, Li H, Hao X, Wang J (2009) Evaluation of ecological vulnerability in the area around Qinghai Lake based on remote sensing and GIS technology. Remote Sens Technol Appl 24(05):635–641
2. Biggs R, Maja S, Michael LS (2015) Principles for building resilience: sustaining ecosystem services in social-ecological systems. Cambridge University Press, Cambridge. https://doi.org/10.1017/CBO9781316014240.
3. Kang H, Tao W, Chang Y, Zhang Y, Xuxiang L, Chen P (2018) A feasible method for the division of ecological vulnerability and its driving forces in Southern Shaanxi. J Clean Prod 205:619–628
4. Li XY, Ma YJ , Xu HY, Wang JH, Zhang DS (2009) Impact of land use and land cover change on environmental degradation in lake Qinghai watershed, northeast Qinghai-Tibet plateau (SCI). Land Degrad Dev 20(1)
5. Lioubimtseva E, Henebry CM (2009) Climate and environmental change in arid central Asia: impacts, vulnerability, and adaptations. J Arid Environ 73(11):963–977
6. Liu J (2011) How to use DEM to make a graded slope map in ArcGIS. Surv Spat Geogr Inf 34(01):139–141

7. Young G, Zavala H, Wandel J, Smit B, Salas S, Jimenez E et al (2010) Vulnerability and adaptation in a dryland community of the Elqui Valley, Chile. Clim Change 98(s1–2):245–276
8. Yu B, Lv C (2011) Evaluation of ecological vulnerability in alpine regions of Qinghai-Tibet plateau. Geogr Res 30(12):2289–2295
9. Zhang, J. (2014). Research on methods and models of ecological vulnerability evaluation in Shanxi Province based on 3S technology. Shanxi Agricultural University.
10. Zhong X, Sun B, Zhao Y, Li J, Zhou X, Wang Y, Qiu Y, Feng L (2011) Evaluation of ecological vulnerability in Yunnan Province based on principal component analysis. J Eco-Environ 20(01):109–113
11. Zhu Q, Zhou W, Jia X, Zhou L, Yu D, Dai C (2019) Ecological vulnerability assessment of Changbai Mountain National Nature Reserve and its surrounding areas. J Appl Ecol 30(05):1633–1641

Printed in the United States
by Baker & Taylor Publisher Services